Weed Management of Horticultural Crops

Weed Management of Horticultural Crops

R.C. Upadhyaya

ANMOL PUBLICATIONS PVT. LTD.
NEW DELHI-110 002 (INDIA)

ANMOL PUBLICATIONS PVT. LTD.
H.O.: 4374/4B, Ansari Road, Darya Ganj,
New Delhi-110 002 (India)
Ph.: 23278000, 23261597
B.O.: No. 1015, Ist Main Road, BSK IIIrd Stage
IIIrd Phase, IIIrd Block
Bangalore - 560 085 (India)
Visit us at: www.anmolpublications.com

Weed Management of Horticultural Crops

First Published, 2008
ISBN 978-81-261-3429-8

PRINTED IN INDIA

Printed at Mehra Offset Press, Delhi.

Contents

Preface

Horticulture has a special significance for the economic development. Most of the people, though hard working and face the calamities of nature and live in harsh conditions, yet produce lot of fruits, vegetables and other flowering and ornamental plants including the medicinal and aromatic plants. In order to increase the productivity of horticultural crops growers must take the advantage of the expertise of scientists. Keeping this idea in focus, the present work has been compiled which contains experiences of the experts and related people. The volumes cover advances in strategies, production, plant protection, value addition and other important areas of horticulture development.

The Encyclopaedia provides practical information on production technologies of tropical and subtropical fruits as well as plantation crops and horticultural nursery. Post-harvest management, marketing and export trends have been furnished for guidance. It also presents most realistic cost estimates and returns from cultivation of such crops. The volumes aim at viewing commercial prospects for horticulture. It has been carefully designed to meet the queries of intending investors, financial institutes, creditors, traders and growers engaged in dealing with horticultural crops. Further it will serve the purpose reference books to students, academicians and related people.

R.C. Upadhyaya

Introduction

Environment and sustainable development issues cut across the whole spectrum of economic activities, and particularly concern agriculture, forestry and fisheries. In spite of their diversity, Small Island developing States share common constraints that impede their efforts to develop the agriculture, forestry and fisheries sectors, while preserving their environment. Among the constraints are fragile ecosystems, vulnerability to natural hazards, population problems, poor communication facilities, and dependence on external economic factors. They run the added potential risk of rising sea level. On the other hand, most islands have the potential to diversify their economies for export and import substitution, for tourism and for drawing greater benefits from the fisheries resources of their exclusive economic zones. In determining their individual sustainable development objectives, Small Island developing States can draw great benefit from improved exchange of information with other islands sharing similar problems.

The agriculture, forestry and fisheries sectors have three vital functions:

- supplying food and raw materials;
- providing employment and income in rural areas; and
- rendering essential environmental services for present and future generations in the maintenance of a large part of the earth's renewable natural resources.

In most countries, these sectors no longer have the capacity to perform all three functions efficiently and simultaneously.

The issues presented in this paper include approaches to sustainable management of the resource base by the agriculture, forestry and fisheries sectors, enhancement of self-reliance through a combination of diversified and intensified primary production from renewable resources, generation of income from the tourist industry, and capacity-building to meet the challenges posed by sustainable agriculture and rural development.

Macro-economic Conditions

More than other countries, small island developing States are dependent on the international trading system for their livelihood, which makes them vulnerable to the vagaries of the world market. This characteristic of small island developing States means that the achievement of sustainable agriculture is more complex than for countries which have to rely less on the world market. The relationship of trade policies to sustainable development and the environment is currently the subject of intense debate.

Fitting into the global economy is especially hard for small island developing States, which face a number of drawbacks:

- limited population size making economies of scale and labour specialization impossible;
- out-migration of skilled personnel, resulting in a high degree of dependence on foreign capital, external assistance and expatriate technical assistance which may not always be sensitive to local conditions;
- a small monetized sector with local private investment limited to low-risk activities such as trade and services; a heavy burden on the public sector faced with limited domestic income-generating capacity reliant on sources such as import duties, fishing rights and remittances from nationals working abroad;
- a narrow resource base creating a dependence on imports of consumer and capital goods;
- and exports limited to primary products from agriculture, forestry, fisheries and mining prone to considerable fluctuations in output and price.

Sustainable development, including food security, nutritional well-being, poverty reduction, and rural development require inputs from many technical disciplines and should be addressed holistically in the planning process. This has been stressed in the structural adjustment programmes. In the provision of policy advice and programming assistance, the main challenge is to assist small island developing States to incorporate environmental and sustainable development considerations into the mainstream of agricultural planning and policy, in line with Agenda 21 recommendations. Sustainable agriculture and rural development (SD) has three essential goals:

- food security through an appropriate and sustainable balance between self-sufficiency and self-reliance;
- employment and income generation in rural areas, particularly to eradicate poverty;
- natural resource conservation and, environmental protection.

"The International Cooperative Programme Framework for Sustainable Agriculture and Rural Development (ICPF/SARD)"

Since 1991, FAO has been streamlining its Special Action Programmes (SAPs) related to sustainable development. SAPs are clusters of projects which correspond to the priority areas indicated in the den Bosch Agenda for Action and subsequently incorporated into the programme areas of UNCED's Agenda 21, Chapter 14 on SARD, with the addition of two specific SAPs on forestry and fisheries as well as a third SAP directly related to the follow-up to the International Conference on Nutrition.

The component SAPs of the ICPF/SARD are as follows: SAPs focusing on policy advice and planning assistance: (i) Country Policy and Programming Assistance for Food Security and Sustainable Agriculture and Rural Development; (ii) Tropical Forests Action Programme; and (iii) Responsible and Sustainable-Fisheries; SAPs focusing on people's welfare: (i) Nutrition and Food Quality; (ii) People's Participation in Rural, Development; and, (iii) Sustainable Development of Rural Households; SAPs promoting the sustainable management of the natural resource base: (i) Land Conservation and Rehabilitation; (ii) Management of Rural Water Resources; (iii) Conservation, Development and

Use of Plant Genetic Resources; and, (iv) Global Programme for Animal Genetic Resources; SAPs dealing with the sound use of agricultural inputs: (i) Integrated Plant Nutrition Systems; and (ii) Integrated Pest Management. Regional and national projects to implement the ICPF/SARD may be formulated for the small island developing States expressing interest.

It will take several years before the SARD process can demonstrate its environmental, social and economic benefits. During the transition period, heavy investments of labour, capital, technology and research will be required to accomplish SARD's goals.

To be successful, SARD must be perceived as a pivot around which the environmental, social and economic fabric of a country evolves. SARD requires a comprehensive approach where all essential elements are brought together within a single framework. Technical assistance and capacity-building should be pursued with all actors involved in agriculture, who have different but complementary roles.

At the national level, sustainable agricultural and rural development would integrate sectoral and cross-sectoral programming assistance in agriculture, forestry, fisheries, poverty alleviation and nutrition; at the community level, people's participation would be enhanced through human resources development and strengthening rural people's organizations; at the household level, all aspects of sustainable management of the resource base, appropriate technology, sound use of inputs and diversification of sources of income would be integrated to increase levels of production.

Economic agents at all levels and in all sectors, including national governments, develop and carry out strategies for sustainable development and management of land, water and biotic resources. This is why all must have access to reliable information and be involved in the formulation and application of national strategies reflecting the recognized congruence of individual and society interest, both short and long-term. The values of conservation and rehabilitation need to be widely publicized and demonstrated, and technical advice and assistance provided to those needing help. Governments can create the conditions and institutions to encourage sustainable production practices by farmers, fisherfolk and other economic agents.

"Agricultural Sector Review in Malta"

In view of assisting the Government in adapting its agricultural policy to the Customs Union and eventual European Union membership, this project (1991-93) reviewed the challenges and opportunities facing the sector. Over the years, Malta has come to depend increasingly on imports of agricultural commodities, mainly from EU countries. This is due to decreasing land resources available for agriculture and rising food demand from the, tourist industry. The, number of factors contributing to trading with EU countries include their proximity but also reduced import tariffs as compared to other countries, the EU policy regarding export subsidies to non-EU countries, the size of the market and facilities for handling and storage of imports. Besides focusing attention on functional improvements of the agricultural support services (notably experimental work, demonstration and extension), particular emphasis was placed on strengthening the capacity of the sector to retain the land base for productive and ecological purposes. The measures recommended include: improvement of public sector management in agriculture and fisheries; improvement in the management of land, fresh water and marine resources; introduction of high-yielding, cost-effective and environmental friendly production techniques; improvement of marketing systems for crop, livestock and fisheries products; a review of the land tenure system; rationalization of trade measures to control imports of agricultural commodities; and a review of price control measures. Financial incentives could be required during the transition period to boost productivity and improve the incomes of full-time farmers.

Implications of the Uruguay Round Agreement

The Agreement on Agriculture of the recently concluded Uruguay Round will have wide-ranging implications for the developing countries, including the small island developing States. Its anticipated impact on small island developing States arises both from expected changes in agricultural commodities markets and from limitations on their freedom to implement policies adapted to local conditions. The former are likely to be contingent and temporary, whereas the latter are structural and long-lasting.

Nearly all small island developing States are net food importers. Because of the reduction in agricultural support in the developed countries, the structural surpluses they generated in the past are likely to disappear. As a result, world market prices will likely rise, especially as developed country exporters are committed to reducing export subsidies. Several small island developing States already faced with serious balance-of-payments problems will find it difficult to maintain their food import volumes without sacrificing other essential imports.

The Agreement contains some safeguard provisions concerning the possible negative effects on least-developed and net food-importing countries. These cover areas such as food aid, technical assistance and short-term assistance in financing normal commercial imports. Whether these provisions would constitute adequate coverage against higher import bills is not clear in view of certain conditionalities for eligibility.

On the other hand, the prices of commodities exported by small island developing States are likely at best to rise only marginally as a result of the Agreement. Indeed, there may be a reduction in the export earnings of countries which have benefited in the past from preferential access to developed country markets. These countries may, however, gain from concessions in non-agricultural sectors. Overall, there is no certainty that they will be net gainers from the Uruguay Round.

The most significant implications of the Agreement relate to the policy environment. Policies allowed under the Agreement are those which have at best minimal production or trade distortion effects. Policies that do not meet this basic criterion may be kept but expenditures on them must be reduced. The least developed countries are not expected to cut support. In small island developing States, maintaining or increasing food production in order to reduce dependence on imports is contingent upon substantial price support. The Agreement implies that this type of support will be more limited in the future. This has implications for minimum safeguards that they may want to establish (in the form of food security stocks or funds) to protect themselves against world market volatility.

It is difficult to make even a broad evaluation of the Agreement on SARD. Policies that can help the environment are allowed under the Agreement, but these can only be implemented in a healthy overall economic climate and may be too costly for poor countries.

Substantial international assistance (both technical and financial) may be needed in the future to facilitate adjustment to the changing macro-economic conditions at a time when: the favourable position in development aid is being eroded, preferential arrangements may be dismantled, opportunities for work abroad are becoming scarcer, local market sizes are limited, the resource range is becoming narrower, and increasing environmental problems and proneness to disasters jeopardize production.

Sustainable Management

Integrated Coastal Area Management

The continental shelves and coastal ecosystems of small island developing States are of major economic significance for settlement, subsistence and commercial agriculture, fisheries and tourism. Demands on coastal resources are endangering the long-term supply of these resources: large parts of coastal areas are being polluted by local or upland resources, fisheries over-exploited and fish habitats degraded, coastal belt mangroves cut, wetlands drained, coral reefs destroyed, fresh water aquifers are subject to depletion and salinity, bio-diversity conservation is threatened, and opportunities for aesthetic enjoyment of scenery reduced. Of particular concern to low-lying small islands is damage to natural systems such as coral reefs, dune lands, and coastal mangrove belts, which significantly reduces the protection they provide against natural disasters of marine origin (*i.e.* tsunamis, tidal surges and wave erosion).

The disruption of small islands societies' traditions sometimes by policy action or strong external factors through accelerated development, demographic growth, high rates of urbanization and often strong emigration fluxes are weakening traditional mechanisms of control, thus rendering coastal resources increasingly vulnerable to over-exploitation. Added to these sources of disruption are new problems such as more frequent natural disasters and rapid environmental changes which result in the degradation of terrestrial and marine habitats in coastal areas.

Increasingly in recent years, it is being understood that the search for solutions to conflicting demands in coastal areas must involve the various users and administrations concerned. With respect to land tenure, historically, small islands featured hierarchies of land ownership rights from which social status was derived. Environmental regulations limiting land use and controlling access to natural resources have been challenged by landowners who interpret any attempt to limit land rights as a challenge to their social status. The involvement and support of traditional leaders in introducing sustainable development practices is also necessary.

By the mid-1980s, the weakness of conventional sectoral planning in reaching sustainable solutions to conflicting demands for coastal resources was apparent. In response, multi-sectoral, integrated approaches to coastal area management began to be developed, in developed countries primarily, but also in some developing ones. UNCED gave support to this process by stressing the importance for small island developing States to develop national policies and management capabilities for multi-sectoral use of coastal areas.

Conservation and sustainable management of resources of coastal areas is known as integrated coastal area management (ICAM). This offers the means to balance the competing demands of different users of the same resources and to manage the resources sustainably.

Most archipelagic developing States have not established systems for the evaluation of their resources, nor the institutional and technical capacities to formulate and implement sound multiple-use strategies. A characteristic common to all small island developing States is the small number of skilled and experienced people available in most fields of activity. Finding people with the necessary experience to staff a government institution responsible for coordinating multiple use of resources, and taking an appropriate role in pro-active management will tax many island administrations.

Coordinating between sectors will be easier if line ministries and users of natural resources recognise their respective roles in the ICAM process and the impact of their activities on resource users in the sector itself and in other sectors. They must also have the capacity to monitor and assess the effect of the activities of other sectors on them.

Though of limited extents, the ecologies of most small islands are complex and varied. Detailed analysis of agro-ecological conditions is therefore required to provide an adequate form of integration and define improved multiple use of the natural resources.

The agriculture, forestry and fisheries sub-sectors are the chief users of resources in the coastal areas of small island developing States. To assist line ministries in their tasks of designing and implementing ICAM, guidelines such as those being prepared by FAO on the integration of agriculture, forestry and fisheries in integrated coastal area management will be most helpful-Guidelines can also be of assistance for planners and resource users at sub-sectoral level. FAO has also prepared guidelines on topics such as the environmental management of coastal aquaculture, the mapping and inventory of mangroves, and coastal rice culture.

FAO considers it important that appropriate emphasis be placed on specific training in integrated management for line agency staff. In this connection, it is collaborating with other UN agencies in the plan for human resources development and capacity-building for the planning and management of coastal and marine areas.

FAO has pioneered the concept of integrated coastal fisheries management (ICFM) as part of ICAM. The basis of this concept is twofold. First, the sea has long been seen as a bottomless sink into which flow pollutants produced on land which degrade marine habitats. Second, due to the open access nature of most fisheries, in many developing countries, small-scale fisheries become the employer of last resort when there is a surplus of unskilled labour in land-based sectors. This results in decreasing yields from already over-stressed fisheries. The open access nature of the resource also facilitates entry of industrial vessels into artisanal fisheries, contributing to overfishing and often resulting in social conflict. In many cases, it is in the interest of the fishery sector, far more than any other, to take a proactive role in initiating integrated management of natural resource use.

"Integrated Coastal Fisheries Management in Trinidad and Tobago"

Within a broader framework of ICAM, the UNDP/FAO project (1993-94) addresses the inter-relationship between fisheries management and other sectors in coastal areas of the Gulf of Paria including: urban development and housing, reclamation of wetlands for agriculture, run-off of agricultural inputs, depletion of natural forests in upland areas, hydrocarbon exploration and processing, and shipping. Data gathering and research for a better scientific understanding of coastal resources, socio-economic and cultural aspects of fishing communities and other users of the coastal and marine areas, greater awareness about the management needs of the fisheries and the coastal area, and the joint action of a number of government (local and national) agencies and users are the main components of this pilot project.

Land and Biotic Resources

Land-use Choices

The small size of many island developing States makes competition for land more severe than in larger countries. Large commercial plantations were frequently established under the colonial system, and continue today in the more fertile plains of island countries, forcing many small farmers to resort to the poorer soils in hilly regions. This, combined with poor land use practices, has resulted in deforestation and soil erosion.

Due to the peculiar geological characteristics of many islands, water resources are scarce, often limited to thin lenses of freshwater floating on sea water, recharged by rainfall. Poor management of watershed catchment areas and of water flowing through agro-ecosystems, particularly irrigation supplies, can lead to water scarcity and pollution of downstream supplies. Population pressure, expanding tourism, and reduced precipitations of recent decades, have resulted in the over-exploitation of groundwater and seawater intrusion, with subsequent further deterioration of water quality and quantity.

Land use planning is a prerequisite for long-term sustainability, since without it, the economic and social fabric of society as a whole will be endangered. Advice on area-specific soil and water conservation, and management practices will be needed for farmers, graziers and others. Such advice will only be adopted if it is perceived as being capable of bringing rapid economic returns to the users without threatening their fragile survival strategies. Success will also depend on guaranteeing long-term rights to land and water resources.

"Land Use Planning, Management and Information System in Grenada"

Grenada is highly dependent on agriculture, deriving 90 percent of its export earnings from this source, though the sector has stagnated over the past three years due to a combination of low export prices and declining productivity of the traditional export crops. The agricultural development strategy of the Government is to improve the productivity of traditional export crops, while diversifying into nontraditional crops for export and for agro-industrial integration into the expanding tourist sector. Emphasis will also be placed on increased domestic food production. The strategy involves bringing idle private and public lands back into production, and strengthening private and public sector institutions. Topography is a key factor in implementing the above strategy.

On the islands of Grenada and Carriacou, approximately 77 percent and 54 percent respectively of the land area has slopes exceeding 20 degrees. Only 33 percent of the soils are deep, well drained, and free of cultivation constraints. The rest, apart from excessive slopes, are shallow or rocky, or have seasonal drainage problems. A favourable feature of Grenadine geography is a well-defined system of watersheds governed by the topographic position of the rivers and streams. The system of watersheds offers the opportunity for the establishment of rational land use consistent with the characteristics and conditions of each watershed, and with the need for soil and water conservation. The main objectives of this newly launched project (1993-94) are to create the technical basis to support rational development and conservation of the country's natural resources, and to develop an effective forestry management policy, and forest area conservation and development programme for the country.

The project will provide specialised technical assistance and practical hands-on training to local technical staff in the establishment and development of computerized natural resources information systems. The project is designed to provide a rapid response to an immediate need affecting the country's food production and the degradation of its natural resources. It is expected to contribute to establishing a sound basis for natural resource planning and policy formulation at both national and watershed levels. More specifically, the following needs to be emphasised:

- sound cropping practices and cropping systems which maintain soil fertility and productivity for sustainable crop yields as well as practices such as soil testing and analysis to determine fertilizer recommendations on a soil-and crop-specific basis. Land evaluation/land use planning are pre-requisites for sound land use and crop diversification. Development of land information systems will make a vital contribution to improved resource management;
- suitable water harvesting methods, both for domestic and agricultural use (supplementary irrigation), aiming at significantly increasing the groundwater recharge as well as collection and storage of rainwater from ground catchments;
- use of biomass residues (*e.g.* bagasse from the sugarcane industry), as a source of plant nutrients, will reduce use of mineral fertilizers in favour of nutrient sources available within the farm. This would contribute to environmental protection and decrease pollution of soils and groundwater.

Land conditions are similar in all coral island countries, and, apart from those affected by human action, soils are homogeneous. The main restrictions on land use are urban encroachment, tourism, established tree crops, especially coconut, or the existence of improved soils devoted to market gardening. Where choices for land use are still available in these islands, integrated coastal area management should be applied in a comprehensive programme of land use planning.

In small low-lying coastal areas, land well suited for agriculture is under severe pressure from expansion of housing, industry, and tourism. Under the market system, with private land ownership, the immediate economic returns from such uses of land always override the longer-term benefits to be derived from sustainable but low-return agriculture.

Economic pressures thus make it difficult for the rural sector to opt for the sustainable use of the resources entrusted to its care, such as:

- light use of coastal mangrove forests, while maintaining the additional functions of nursery areas for marine species, and protection against major damage to adjacent land from cyclones or tidal waves;
- brackish-water fish ponds in humid climates or salt pans in and or semi-arid conditions;
- cultivation of wetland rice or appropriate dryland crops, irrigated or rainfed, with flood protection and water control (mainly drainage facilities);
- cultivation of annual crops, adaptable to changing conditions, or perennial crops, entailing long-term decisions and larger investment; and
- conservation of local germplasm offered by traditional farming systems.

Hilly or mountainous areas face a different, but no less difficult, set of choices. Depending on climate, geology, landform and soils, the land may have high or low potential for nature conservation, forestry, grazing, production of perennial or long-term crops or cultivation of annual crops. Conditions can vary greatly over short distances, and the optimal response to a country's needs and objectives may well be a mosaic of different land uses and protected areas.

In small island developing States, the relationship between forests and trees, and other types of land use and economic activities, is critical. Forestry should be systematically represented in national and district planning commissions, land use zoning activities, town planning, energy and infrastructure strategies, as well as in planning for generation of income, employment, and food security. Forests and trees and the products and services they provide must be validated and valorized in view of their role in the protection of water resources and of the environment in general.

"Tropical Forests Action Programme"

Forest resources contribute to environmental protection and food security. Rural people exercise important customary rights over forest products, which have traditionally been used sustainably. Current forestry problems originate, among others, from official policies which are not supportive of sustainable management and utilization of forest and tree resources, The importance of forest resources has motivated many Caribbean island countries to develop appropriate inter-island forestry and environmental networks to promote forestry research, as well as carrying out Tropical Forests Action Programmes (TFAP) in order to implement the sustainable management of forest resources. The TFAF planning phase in Antigua and Barbuda, Barbados, Dominica, Grenada, Montserrat, St. Kitts and Nevis, St. Lucia, St. Vincent and the Grenadines, and Trinidad and Tobago is concluded and the implementation started in November 1993. This multinational exercise defined regional issues and produced key documents, including a survey of forest management and silvicultural systems in the region.

In the Pacific, Fiji is already implementing a TFAP-framed national forestry action plan, Vanuatu and Solomon Islands are starting the planning phase, and others (Kiribati, Micronesia, Tonga, Tuvalu, and Western Samoa) have expressed interest in launching a TFAP process. A TFAP Training Workshop for this region is scheduled for May 1994. In Africa, Cape Verde Islands are in active preparation of their National Forestry

Action Plan, and a recently held national Forestry Congress has highlighted the overall progress achieved and original aspects of forestry in the country. Legislation and planning exercises are under way in the Comoros Islands.

Watershed Management

Urgent implementation of watershed management and rehabilitation programmes in most small island countries is needed for sustainable development. On most small islands, distances between highlands and coastal areas tend to be short. Under such conditions, the role of forest ecosystems as regulators of water supplies for the island crop environment and soil formation and protection is especially critical. By extension, the sustainability of agricultural productivity and potential greatly depends on the regulatory functions of forests.

Protecting forests and trees is essential for warding off environmental degradation and rural poverty. The multiple goods and services available from forests offer many possibilities for off-farm employment and for economic diversification, with low capital input, for example through the integration of plantations with agricultural crops, in soil conservation and watershed management, as recreational areas, and as a source of subsistence.

In spite of the significance of forests and tree-based resources, present trends are not encouraging; forest resources continue to be poorly managed and not used rationally in many island countries. Deforestation occurs due to clearing for agriculture in areas not suitable for this purpose, culling for fuelwood, and urban and infrastructure development. Population growth and economic reforms pose new and increasing demands on the forestry sector. The result is accelerated loss of forest cover and of its associated benefits.

"Watershed Management and Conservation Education in Western Samoa"

This UNDP/FAO project (1992-95) looks at the inter-relation between the management of upstream and downstream areas within a given watershed. This has been shown through the implementation of soil conservation and watershed management measures such as conservation plantation, agroforestry practices, plantation of forest and fruit trees by farmers, conservation farming systems, extension and education programmes for different target groups. A monitoring system has also been developed to increase understanding of the impact of upstream management on water quantity and quality and other downstream values.

The destruction of vegetative cover by frequent tropical cyclones has led to accelerated soil erosion and landslides, with a direct impact on the watersheds, agricultural production, water supplies, hydropower generation, lagoons and reefs. The rehabilitation of upstream areas and the participation of the local population in conservation of downstream areas an integral parts of the project. This will facilitate the conservation of flora and fauna and improve the socio-economic condition: of watershed dwellers.

Greening Activities

Considering the limited land area and the islands' comparative fragility, the role of vegetation takes on particular importance. Some islands have no land suitable for agricultural activities and are confined to producing vegetables and growing fruit trees. The cover of low-lying islands is composed of mangroves and other plant associations, as well as exotic species. Demands for boats, construction material, rural housing, resort development and fuelwood have resulted in depletion of tree cover. This has affected the dynamic influence of ocean, coral reefs, land formations, and vegetation on each other.

Drought and desertification in some small island developing States renders afforestation difficult. In order not to interfere with endemic species and corresponding life systems by introducing exotic plants, it is preferable to use endemic species (bringing them in from nearby islands if local vegetation has been degraded) when undertaking forestry plantation and afforestation. Current water harvesting techniques, including soil manipulation devices aimed at concentrating rain water to better benefit planted seedlings, make plantation efforts in dry environments less uncertain than hitherto.

Further research is necessary for:

* improved knowledge of the original indigenous plant species and formations;
* selection and use of local species suitable for incorporation in soil and water conservation, range regeneration and fodder production, and fuelwood supply; and
* screening and eventual introduction of suitable non-indigenous species tried under similar conditions for use in all the activities mentioned above.

Greening activities should also include the identification, protection and proper management of natural formations.

"Development of Forest Resources in Cape Verde"

Within the national development framework, this project (1984-93) offered institutional support to the Government for the establishment and consolidation of an efficient forestry administration and the creation and management of enhanced forests and rangelands to promote rural development. These activities will contribute to combating, desertification, rehabilitate the environment through the reconstitution and conservation of the vegetative cover, establish integrated production system, contribute to the provision of wood for multiple uses (energy, construction, etc.), enhance socioeconomic welfare through the sustainable exploitation of forest resources, and transfer management responsibilities to a competent administration working within an adequate institutional framework. This project, completed in December 1993, should 'facilitate, *inter alia*, the finalization of the National Forest Action Plan, leading to better coordination in the planning and implementation of sector policies while promoting the participation of rural people in the preparation and management of the action plan.

Protected Areas and Biological Diversity

Island bio-geography has led to a high endemism in terrestrial species, especially in larger islands (*e.g.* over 1,000 indigenous species in Fiji). All small island developing States are particularly vulnerable to the impact of introduced species and deleterious human action.

The physical characteristics, the biotic formations, and socio-economic situation of small island developing States determine the conditions and role of forestry in sustainable development. The implications of their small size, their isolation, and the nature of their forest resources define the available models and opportunities for forest development, and suggest the necessary precautions to be taken in any manipulation of plant and animal populations. The increasingly negative impacts on local economies and on the quality of the environment are already visible in many cases through shortage of water supply, soil erosion, increased sedimentation of rivers, estuaries and coastal areas, and coastal erosion. Deforestation and forest degradation have led to the extinction of many animal and plant species, causing irreversible loss of genetic resources and ecosystems. The establishment of protected areas, such as forest reserves, national parks and wildlife sanctuaries, supported by botanical gardens, herbaria, etc., are essential for conservation of biological diversity in these countries. Those areas can become the basis for eco-tourism.

Marine ecosystems are prone to destruction of habitats and bio-diversity. There are several causes of this, including: destruction of coral reefs (both deliberately by fishermen or tourists and indirectly by pollution, sedimentation, and land reclamation for building sites); conversion of mangroves and wetlands (resulting in loss of important nursery areas); the use of long driftnets (mostly impacting marine mammals, turtles, and non-targeted fish) by a growing number of foreign fishing vessels; and more generally over-fishing. Lack of monitoring makes it difficult to quantify the damage of such activities on marine life. Surveys are needed to identify representative habitats and ecosystems for the establishment of marine reserves and parks.

Plant Genetic Resources

Plant genetic resources are both the building blocks of all living matter, and the raw material for the fast growing plant breeding and biotechnology industries. According to recent studies, any region of the world is dependent on genetic material which originated in other regions for over 50 percent of its basic food production; and for small island developing States, it may be close to 100 percent.

The conservation *(in situ* and *ex situ)* as well as utilization and management of plant genetic resources require both financial and human resources that may not always be available in small island developing States. Cooperation between those States is therefore desirable in this area to safeguard these resources for future use. It is of particular interest for small countries to have access to genetic resources from other countries in the same agro-ecological zone. Means are needed for conservation and evaluation of existing basic plant genetic resources, and for capacity-building for evaluation of crop cultivars for practical farming. The latter because sustainable farming systems need access to crop cultivars with an optimal production potential, good resistance and quality characteristics, and adapted to local environments.

Much of the plant genetic diversity still in active use in developing countries is maintained by small farmer-breeders as part of their livelihood and personal use. FAO has developed, within its International Undertaking on Plant Genetic Resources, the concept of Farmers Rights which recognizes the contribution of generations of rural people to domestication of wild species and their subsequent improvement through on-farm breeding. Farmers have thus conserved germplasm and made it available to societies and continue to do so. While it would be utopic for them to claim individual compensation, the concept form the basis of formal recognition and reward system to encourage and enhance the continued role of farmers and rural communities in the conservation and use of plant genetic resources. This aims at reconciling the view of the "technology rich" and the "gene rich" countries in order to ensure the availability of plant genetic resources within an equitable system.

"Improved Seed Production in the Caribbean"

Production of good quality seed is very limited in the Caribbean. With the exception of sugarcane, there is practically no germplasm information or exchange of species and variety performance among countries. In general, countries lack the infrastructure and technical skills to produce the required seed.

Selected seeds (mainly vegetables) are imported, in most cases without any quality control or adaptation trials this results in frequent unsatisfactory germination and low yields. This project (1992-94) is developing a regional seed technology training programme and an international information exchange network on plant genetic resources availability, variety descriptions and variety trails. Field and laboratory standards are established for seed quality control for both sexual seed and vegetative planting materials, to facilitate active germplasm and variety exchange programmes among countries (Antigua and Barbuda, Bahamas, Barbados, Belize, Dominica, Grenada, Guyana, Jamaica, Montserrat, St. Kitts and Nevis, St. Lucia, St. Vincent, Surinam, and Trinidad and Tobago). Specific seed activities will be supported in selected countries. It is intended in a second phase to promote active germplasm exchange, seed production, and further development of seed technology.

The concept of Farmers Rights also provides the counterbalance to "formal" Intellectual Property Rights and Patents which compensate for the latest innovations (technologies and/or others) without giving enough consideration to the fact that in many cases, these innovations are only the last step of accumulative knowledge and innovations carried out through many generations over millennia in different parts of the world.

Small farmer-breeders are generally beyond the reach of the seed distribution infrastructure and maintain actively considerable genetic diversity. The Convention on Biological Diversity has left-out two outstanding issues: access to ex-situ collections and farmers rights. In order to address these issues, FAO has initiated a countries' negotiating process to revise the International Undertaking on Plant Genetic Resources which would be presented to the Conference of the Parties on the Convention on Biological Diversity as a possible protocol.

During the late 1980s, FAO assisted the countries of the South Pacific region for collection, conservation, characterization and documentation of root and tuber crop germplasm. Currently most of the collections are held in the Dodo Creek Research Station, Honiara, Solomon Islands. In the coming biennium,

FAO proposes to survey the Pacific region to evaluate the status of existing *ex situ* collections and to propose action to address any related problems and strengthen national capabilities for conservation and sustainable utilization of existing germplasm collections.

Managing Risk due to Natural Hazards

Cyclones inevitably affect any programme dealing with sustainable agriculture and rural development. Early-warning systems and disaster preparedness, benefiting from the collaboration of several international organizations to minimize the impact of cyclones, are already under operation in many countries.

The effects of hurricanes and cyclones on forests and forestry plantations are a major constraint on sustainable forestry development in the Pacific, Indian Ocean and Caribbean islands.

Research and development into forestry and natural resource conservation and management are needed to mitigate the effects of strong winds and cyclones on these countries, in particular:

* salvage logging after windblow, including inventory, logging techniques and the storage of large quantities of logs before processing;
* selection of species and tree breeding through hybrids or other improved material of wind-firm trees;
* tree breeding strategies where trials or seed trees are liable to be destroyed; and
* management techniques especially related to reduction of damage through initial spacing, pruning and thinning.

There is thus a strong need to develop technology-transfer programmes to promote environmentally sound forest operations which take account of the special environmental risks associated with small island developing States.

"High Wind-resistant Trees"

In 1991, Cyclone Val destroyed about 60 percent of the vegetative cover in the Vaisigano watershed area in Samoa. The result was accelerated soil erosion and landslides, and increased surface run-off. Agricultural production, water supplies for consumption and hydropower generation, lagoons and reefs were all affected. As a quick rehabilitation-measure, the area has been planted with *Albizzia* species at the spacing of 10 x 5 metres to allow the regeneration of natural vegetative cover.

This tree species was selected because of its resistance to high winds (even when badly damaged by cyclones. *Albizzia* revives), low-labour requirements (being a fast growing species, no maintenance is required after 1.5 years), capacity to improve soil fertility and ability for self-regeneration within 4-5 years (thus providing quick cover). In Fiji, the pine *(Pinus Caribaea)* forests are liable to exposure to storm or hurricane force winds at least twice during a pine rotation. Wind damage to the forests is often severe and occasionally devastating. Research over several years has aimed at improving crop stability in plantations. Achievements include the establishment of clonal seed orchards introducing a group of more windfirm coastal provenances of pines selected from survivors of hurricane force winds. Species trials have shown that several hardwood species have greater vigour and stability than *P. caribaea* and may be valuable as shelterbelts on very exposed sites.

Apart from the direct impact of cyclones on crop production, the major risk to agriculture is salt-water intrusion on both soil and freshwater resources. Strong storms can result in complete coverage of low-lying islands by waves, preventing subsequent growth of crops for many months due to residual salt. Most importantly, salt water contaminates the limited freshwater lense underlying atoll islands which is the only source of water for both drinking and agricultural purposes.

The gravity of salt-water intrusion into the freshwater lense would he exacerbated if the predicted sea-level rise materialises: lateral leakages would increase, lenses would become thinner, salt-water would reach water-pumps and wells, and the roots of tree crops such as coconut palms and breadfruit.

Agricultural insurance is available for weather damage and the basic requirements for this are well documented in FAO. The Organization is in a good position to provide advice on the management of the attendant risks, particularly adverse weather events, such as cyclones.

"Banana Insurance in the Windward Islands"

Bananas are the principal export crop of the Windward Islands (St. Lucia, Dominica, Grenada, St. Vincent). The crop is vulnerable to excessive winds. Individual growers, (totalling some 25,000) who export through the group marketing scheme, WINBAN, are compulsorily covered under an insurance scheme run essentially on commercial lines, and embracing all four island nations of the Windward Group. The premium rate charged is actuarially determined and is currently approximately 13 percent of the sum insured. The scheme is entirely funded by the premium income. Given the catastrophic nature of hurricanes and the possibility of massive losses, re-insurance is arranged by WINCROP on the international market. Loss assessment is carried out by WINCROP staff, assisted by on-call assessors, as required. Records for the entire operation are computerised, which means that loss assessors are provided with a computer printout giving details of each banana operation-area, past production, past loss assessment-a key to enabling this scheme to run efficiently, without subsidy.

Rural Energy Resources

People living on small islands frequently face difficulties in maintaining a basic standard of living; isolation, high transportation costs, limited markets for services and goods and the lack of indigenous resources are factors that affect every aspect of life. Most marketed goods are expensive or difficult to obtain, not least a reliable energy supply or domestic fuel.

In most islands, the better-off people enjoy the range of household or industrial facilities common to cities and towns everywhere. Urban communities on most islands depend upon a municipal electricity supply. Power is normally generated centrally from oil or gas delivered to the island and, increasingly, from indigenous resources such as agricultural residues.

Renewable sources of energy are sometimes available, including geothermal sources, wind generators and hydro-electric turbines, but they are relatively limited compared to conventional generation (usually the internal combustion engine), mainly because the costs of exploitation and use normally makes them uneconomical compared to imported fuels.

Notwithstanding the dependency that this implies, and the need to obtain foreign currency with which to purchase oil or gas, there are significant advantages to utilizing commercial sources of power and equipment. Liquid or gas fuels are versatile and compact, and the range of services and equipment available with which to use them is extensive, at least for urban dwellers.

Providing electricity to rural communities is, however, costly and this may offer opportunities for developing renewable sources of power. Currently, the range of equipment and services for using alternative energy resources is limited, often expensive and sometimes difficult to obtain. Moreover, much of the equipment is site-specific, for example, wind powered equipment.

Government policies can encourage the introduction of renewable sources of energy. This has been the case with wind-generators in Denmark during the past 20 years. The use of renewable sources of energy, though sometimes expensive in terms of direct costs, can be more easy to justify on environmental grounds.

Agricultural and forest biomass is the oldest form of fuel and remains the most common one for rural and urban households the world over, whether for heating and cooking or for cottage industries. Fuelwood sometimes constitutes an important source of income for those living around forest areas. For densely populated islands, exploitation of wood for fuel is no longer a viable option and dependence on imported fuels has generally been on the increase.

Wood for energy can be grown in plantations or woodlots, or derived from sustained management of forest woodland, agroforestry or trees outside the forest. This constitutes an important potential source of fuel compatible with conservation of the environment and is complementary to other important functions of forests and trees and associated housing areas. Bagasse as a feed for ruminants is well-documented and sometimes widely used. Household cooking stoves fuelled by rice husks have been developed and are popular in South East Asia, but rarely seen in other rice-growing areas.

In many small island developing States, animal traction has an extremely important role to play as a source of energy. In the Dominican Republic, La Romana Sugar Estate, one of the biggest sugar factories in the world, has kept up to 18,000 oxen for transporting its sugarcane to the factory. Another increasing contribution of animals to energy production comes from biogas. New simple biodigesters made of PVC plastic sheets are now used in various tropical countries to produce cheap energy for use in kitchens.

The role of FAO, and the other international agencies, in the field of energy resources for island developing States depends in the first place on requests from governments for assistance. Such requests are few, perhaps due to the aggressive policies of the commercial power generation industry, particularly for petroleum. Renewable power generation industries also exist, but they are smaller and their capacity to influence policy is limited.

There is scope for providing information, contacts, companies, case-studies, etc. to enable people to compare and contrast experiences, and to obtain technical and commercial information. FAO can also assist with the preparation and execution of projects which focus upon energy supplies. These can cover a wide variety of activities, from fuelwood plantations to solar-powered telecommunications or cold-storage for food or medicines.

Macro-planning at national level, to assist governments (and the private sector), to determine longer-term strategic energy supplies is well within FAO's technical resources. However, so long as price is the sole criterion in determining energy supplies and environmental costs are not included in price calculations, alternative, renewable resources are unlikely to gain a larger share of the energy market.

"Bio-energy from Agricultural Residues"

A large number of small island developing States, mainly in the Caribbean and the Pacific, are major sugar producers and this agro-industrial activity remain their main source of income. Sugar mills are large consumers of energy and have long burnt sugarcane bagasse to generate the electricity and heat required in the production process. Depending on the efficiency of the system, normally directly related to the age of the plant, these mills can be self-sufficient in energy terms, and can even produce extra power to feed to the electricity grid.

The amount of bagasse produced usually far surpasses the requirements of the mills, and in many cases, the disposal of bagasse is a major environmental problem. Many island States have embarked on a process of modernizing their sugar mills in order to improve their economic operation. Among the changes being introduced are better combustion and larger power generation systems.

The potential of these bioenergy resources is considerable. Production of ethanol as fuel, of animal feed and of fibre are among the potential uses and this could open new economic and employment opportunities to small island developing States. Nevertheless, technical and legislative measures need to be developed and implemented to overcome barriers to the full use of the energy potential of these biomass resources. In many countries, legislation still does not permit power generation and especially power feeding into the national grid by institutions outside the traditional power sector. Environmental policies will undoubtedly point towards the promotion of such schemes.

With moderate inputs and suitable management, existing varieties of sugarcane produce reasonable yields in marginal lands opening up new energy and environmental opportunities. FAO organized in 1993 an Expert Consultation on Biofuels for Development and plans to expand its activities in this: field through global and national studies. Biomass resources can contribute to raising incomes in the agricultural sector, and to the achievement of stable and self-sustainable energy systems.

Edibility of Ornamental Tree Fruit

Ornamental trees, such as crabapples and purple-leaved plums, are planted in the home landscape for their flowers or colourful foliage. While chiefly ornamental, these trees also produce fruit resembling those on fruit trees. As the fruits mature, questions concerning their edibility often arise.

Fruit from ornamentals, such as crabapples and purple-leaved plums, are edible. However, the eating quality of some ornamental fruit is rather poor. These trees were selected for their ornamental characteristics rather than the eating quality of their fruit.

The crabapple (*Malus spp.*) is an example of a common ornamental that also produces edible fruit. Apples and crabapples are differentiated strictly on the size of their fruit. Crabapples are defined as those varieties with fruit 2 inches or less in diameter. Those with larger fruit are apples. At maturity, crabapple fruits vary from yellow to orange to bright red. The fruit of some crabapple varieties colour and ripen in August, others mature in the fall. (The colourful, persistent fruit of many of the newer crabapple varieties are actually an important ornamental characteristic.) While all crabapple fruit can be used in making jellies and preserves, large-fruited varieties, such as 'Whitney' and 'Chestnut,' are the best.

Purple-leaved plums (*Prunus cerasifera*), such as 'Newport,' 'Thundercloud,' and 'Mount Saint Helens' are planted in the home landscape for their white or pink flowers and purple foliage. In most years, ornamental plums produce little or no fruit as they bloom so early that pollination is generally poor. Occasionally, however, they produce a good crop of fruit. At maturity, the edible fruits are approximately 1 inch across and dark purple in colour.

The new growth of 'Canada Red' and 'Schubert' chokecherry (*Prunus virginiana*) emerges green, but quickly turns to a maroon red. The small trees produce clusters of white flowers in spring followed by round, 1/3-inch diameter fruit. At maturity, the fruit are dark purple to black. Though bitter, the mature fruit can be used for jams, jellies, and juices.

'Autumn Blaze,' 'Chanticleer,' and 'Redspire' are ornamental pears (*Pyrus calleryana*) grown for their glossy dark green foliage and small white flowers in early spring. Ornamental pears generally produce only small numbers of fruit. The inedible fruit are roundish, 1/3-inch across, and tan to brown in colour. A word of caution. If you are uncertain of the identity of a tree or shrub, don't eat the fruit. It's better to be safe than sorry.

Incredible, Edible, Ornamental Sweet Peppers

Think of a vegetable whose fruits can be green, oranges, red, gold, lime or even purple. Combine this broad colour spectrum with crunchy texture and mellow sweet taste, endless culinary uses, ease of growth and compact plant habit, and you've got a versatile ornamental vegetable — the sweet pepper.

From the classic bell type (referred to as a "mango" in some areas) to its relatives the 'Cubanelle,' pimiento, 'Sweet Banana', 'Sweet Cherry' and sweet wax types, sweet peppers offer a palette of culinary and visual pizzazz.

Sweet peppers are certainly one of the premier edible ornamentals. In recent years they've emerged from the vegetable patch and are now found lending their decorative, festive touch to flower borders, patio gardens, entrance yards, porch boxes, containers, and all types of locations calling for productive, colourful plants.

Even the most common pepper type, a green bell such as 'California Wonder,' provides a visual delight. Much to the surprise of many novice gardeners, the normally shiny green fruits develop further to brilliant red when left on the plant to full maturity. A check at the grocery produce counter shows that not only the colour changes, but the fruits' value goes up, often by a factor of two to three times. Red bells are also far sweeter and more nutritious than green ones, and their mellow taste makes the wait for maturity worthwhile.

Pickling Favourites

Look to the smaller-fruited sweet peppers for more variety and sheer numbers of fruits. Types such as 'Sweet Cherry' and 'Hungarian Sweet Wax' consistently produce three dozen or more fruits per plant. The golf ball size fruits of 'Sweet Cherry' are shiny green at first and finally turn brilliant red, making them perfect for relish trays or salads. Their shape and small size make them favourites for pickling whole. A single 'Sweet Cherry' plant is tailor-made for growing in a three-gallon container, or for combining in a larger container with other vegetables and flowers. The sturdy, bushy plant seldom grows taller than 20 inches and sparkles with its crop of bright fruits.

'Hungarian Sweet Wax' is another favourite for pickling. Its 5-to 6-inch long tapered fruits come in a mix of yellow, orange and red, depending on degree of maturity. They pack easily into quartsize canning jars, making an attractive display.

Watch seed catalog descriptions and bedding plant labels carefully because you don't want to confuse 'Hungarian Sweet Wax' with its look-alike, but fiery hot cousin usually sold as 'Hungarian Wax' or 'Hungarian Yellow Wax.' Most seed catalogs list hot varieties separately, but at bedding plant outlets, labels are sometimes inadvertently switched.

'Sweet Banana' is a prolific producer of fruits with size and shape similar to those of 'Hungarian Sweet Wax.' Colours range from lime-green through yellow, orange and finally red. A pickling favourite, 'Sweet Banana' is as good as nearly any pepper for fresh use, and is widely used for frying. 'Top Banana,' a new hybrid introduction this spring, is described as "outclassing open-pollinated varieties with prolific production and much better performance under stress." 'Top Banana' matures its first fruits in about 63 days from transplanting, making it a likely choice for short season areas.

Few varieties beat 'Cubanelle' when fried peppers are on the menu. The pendent, tapered fruits usually reach a length of 6 inches and measure 2 inches or slightly more at the shoulder. 'Cubanelle' is best harvested at the yellow-green stage, when the fruit walls are still firm and hold up well. Fruits turn red and become sweeter at full maturity, but lack substance at this stage.

In Northern Zones

Gardeners in short season areas can expect best production from early maturing varieties. Assist these carefully selected types with season extenders such as cloches or similar covers, and heat-trapping mulches. Utilize a planting area facing a south or west wall.

This may make the difference between success or failure in a marginal climate. Plants growing in containers have the extra dimension of portability, permitting you to move them to ideal locations for growth, and later to a spot where their colourful, edible fruits can be enjoyed.

As you plan this year's garden, keep the ornamental versatility of sweet peppers in mind. Give them a try in a few locations beyond the vegetable garden. You'll be rewarded with easy-to-maintain ornamental plants, and abundant harvests of attractive, nutritious and delicious fruits.

Global Horticultural Impact

Since 1985 horticultural produce and commodities have been evincing increasing importance in developing countries, which have increased their overall share of world production and now have fruits and vegetables making up a large proportion of their exports. Future growth will likely depend on rising and changes in demand in both developed and developing countries and on the capability of emerging economies to maintain or increase their competitive strength in world markets. The implementation of the Uruguay Agreement will impact on trade volumes and trade flows particularly through such general areas as market access, tarification, limits on export subsidies, cuts in domestic supports, phytosanitary measures and safeguard clauses. The brief of the present study is to gain a better understanding of the economic significance of, and outlook for, production, consumption, imports and exports of fruits and vegetables, and key issues related thereto, in developing countries.

In recent years analysts have been taking a much closer look at the horticultural sector of developing countries, especially in regard to its potential contribution to income generation and export earnings (OECD, 1996; Islam, 1990). The literature is full of specific studies on selected horticultural produce/countries (Segre *et al.*, 1997), although little research has been done to arrive at a comprehensive analysis of production, consumption and trade. One factor explaining the lack of such a comparative approach is the limited range of available data.

On the strength of a homogeneous FAO-derived worldwide database (World Tables of Fruits and Vegetables Development, WTF&VD) specifically worked up for its brief, the present paper adds to this small fund of knowledge so that a better understanding of the economic significance of and outlook for production, consumption, imports and exports of fruits and vegetables (excluding such products as cut flowers, foliage, potted plants, and nursery materials) in developing countries can be achieved. The data reported in the following pages cover 61 products over the period 1985-1996 for 168 countries. The following sections also summarize certain findings that were part of a wider analysis conducted in the framework of the WCHR-Working Group 1. The complete information set, including a widely cast bibliographic net, can be found a widely cast bibliographic net, can be found in the working document by Segre et al. (1997), while the summary tables of the database are now on the WCHR Web Site (Segre et al., 1998) and the country-by-country and product-by-product dataset will be available upon request to the author.

Methodology

The present study combines statistical and economic analyses. The former is performed by the WTF&VD. These tables, derived from the FAO database, cover the time period 1985-1996 and consider some 228 countries arranged in 17 country groupings. Only the aggregate "Developing Countries" (hereinafter DCs) is considered here, which enumerates nine country groupings: 1. North Africa, 2. Sub-Saharan Africa, 3. Central America and Caribbean, 4. South America, 5. China, 6. Near East in Asia, 7. East and South-East Asia, 8. South Asia, 9. Oceania Developing. Within each one three product groups are examined-vegetables, fresh fruits and other fruits-and for each such group in turn the following indicators are reported: production, imports and exports (in volume and value). Five-year averages, annual growth rates and shares were than calculated, and total and per capita (apparent) consumption and a standardized trade balance index were derived. It should be noted that in the WTF&VD all these figures are available also at comp; VD all these figures are available also at country and product levels.

These data were then used to apply an economic model known in the literature as *portfolio analysis* (Valdani, 1992), which by crossing growth rates, market shares and averages of production, exports and imports focuses on the dynamics of competitiveness in fruits and vegetables between the chosen countries/country groupings.

The results of past trends and salient characteristics of fruit and vegetables in DCs, as based on the aggregate data *supra*, raise a number of issues and questions-strengths, weaknesses and advantages for production and trade, possible impact of the general trend towards trade liberalization, implication of export competition, and the like. The statistical and economic analyses are therefore complemented with more detailed examinations of specific related issues, especially experiences of an individual country/product in the area under investigation. Concluding remarks and future prospects are reported in the final section *infra*.

Main Results and Related Issues

This section reports the main findings of the statistical and economic analyses. For reasons of space, only the main aggregates are considered, *i.e.* vegetables and fresh fruits and country groupings. Some comparisons with the aggreand country groupings. Some comparisons with the aggregate "World" (developing countries plus 8 areas of developed countries) are also presented for comparison. Comments on the trends and related issues are kept to the essential.

Vegetable Production

DCs account for 72% of worldwide vegetable production (1996). Over the time period 1986-95 production increased, on average, at a growth rate of 4.11%, a faster pace than the 2.11% world trend. Total production surged from 263 million metric tonnes (MMT) in 1986-90 to 327 MMT in 1991-95. All nine examined areas showed a positive trend, although the 6.6% growth rate for China was much higher than that for the other country groupings. By 1996 China accounted for more than 48% of the total vegetable production in DCs, a more than 10% rise from its 38% in 1986. South Asia was by 1996 the second leading producer at 19.4% of total vegetable output, followed by the Near East (8.48%) and East and South-East Asia (9.14%).

Of particular note here is the importance of China's top ranking in this cere is the importance of China's top ranking in this context and the impact it now has on the world horticultural sector as supplier, especially in the light of China's limits in storing and processing its domestic production. Among the other DCs, the developing vegetable production raises two priority issues, one related to income generation and the other to introduction of new types of vegetables.

Income Heneration

The evolution of vegetable production has an important economic significance in this area as output is relatively labour and capital intensive and vegetable growing by farmers is consequently dependent upon whether returns can at least cover production outlays. The role of vegetables in income generation is best illustrated through country examples.

In Vietnam much of the increased demand for fresh vegetables from the urban population, particularly for the more perishable species, is being met by peri-urban production, which has significantly increased since the change from a centrally planned economy with collective systems to a market-orientated one. Vegetables provide about $650 value added (returns to labour, land and management) per farm yearly for peri-urban vegetable farmers.

Guatemala provides a similar example. Vegetable production may be a promising option for meeting the two national goals of crop diversification and expansion of non-traditional agricultural exports because of rising foreign demand. Targeted to the smallholder sector, such production creates opportunities for labour-absorbing production techniques, although it also exposes farmers to greater risks related to market uncertainties. Production of export crops by small-holder farms in Guatemala appears to have had a positive effect on household income and food security, particularly for the smallest farms. Production of export vegetables increased on-farm employment, reducing the need to rely on uncertain off-farm employment in local and distant labour markets for additional income. New employment opportunities were also created for non-cooperative-member households. Disparities in income levels were also reduced (Von Braun, 1994).

Also in Malawi a wide range of exotic and indigenous vegetables are grown that play a critical role in the nutrition of the people by providing essential minerals and vitamins and generate some income for small-holders. This has prompted the Ministry of Agriculture more vigorously to pursue research, extension and personnely to pursue research, extension and personnel training in vegetable production (Mkamanga, 1990).

Another interesting example is in the North Indian state of Himachal Pradesh, where returns from vegetable growing are quite high and could be much higher if the package of recommended practices is adopted. Increases in vegetable acreage on small and marginal farms will not only provide gainful employment for under-utilized family labour but would also reduce income disparity among farms of differing sizes (Bathi, 1993).

New Types of Vegetables

The future of the vegetable industry in DCs will necessarily depend on improving people's standard of living as well as on the increasing, and more sophisticated, demand of consumers in large towns and cities for produce enhancement in terms of colour, taste, smell and shape. The new types of produce these countries, and China in particular, want to develop or to introduce from foreign countries are spicy vegetables, vegetable powder and convenience vegetables.

The demand for spicy vegetables depends on the fact that many people have become tired of mediocre delicacies but pungent spicy foods seem to be well-received all the time. Some food specialists believe that this type of produce can help dissolve fat and reduce its sediment in human tissues. Spicy foods are also said to be cancer human tissues. Spicy foods are also said to be cancer-resistant. Most vegetables can be reduced to powder after they are washed clean, the usual method being to pulverize, sterilize, dehydrate, dry and package (in plastic bags) them. The powder can be added to fermented dough before being processed into vegetable biscuits or noodles. It could also be made into instant vegetable soup, especially catering to the needs of those children or infants who dislike fresh vegetables. According to some food specialists, these vegetables show rosy prospects in sales and have become so popular in some big cities in the past two years that supply has fallen short of demand (USDA, 1998).

Packaged vegetables with a registered trade mark are growing quickly. The appearance of this type of produce in the marketplace indicates that many DCs will industrialize the mass production and processing of vegetables and establish a number of domestic and global name brands of their own. These products are thought to have a very bright future in the coming 21st century.

Production of Fresh Fruits

In 1996 DCs represented some 72% of World fresh fruit production. As with vegetables, China registered the fastest growth rate (7.61%) for fresh fruit production in all DCs, wherte (7.61%) for fresh fruit production in all DCs, where the average growth rate over the period 1986-95 was 3.25% (World: 1.46%). Production climbed from 253 MT in 1986-90 to 302 MT in 1991-95. Together with South America, China also garnered the largest market share with one-fifth (1996) of total fresh fruit production, the remainder being equally distributed in the other country groupings. China has a potential very similar to South America's, but with a much higher growth rate. The difference between the two areas consists in the fact that while South America is export-oriented towards the US and the EU, Chinese production is still earmarked for local consumption.

These trends raise at least two issues-the importance of the quality and stability of supply (China) and the role played by agricultural policy reforms (South America).

Quality and Stability of Supply

Fruit supply, in effect, should offer a minimum standard of quality and stability. Buyer surveys have clearly established the importance of quality, reliability of delivery and price as determining factors. Quality elements comprise grade of produce, defects, ingredients, mode of preparation of raw materials for processing, preparation and density of liquerials for processing, preparation and density of liquid medium, additives, drained weight, cleanliness, strength and fill of container and packaging material, and storage conditions to safeguard quality retention.

Although the fruit handling systems and marketing arrangements in DCs have evolved to take account of environmental factors and the lack of appropriate support systems, local marketing of fruit is still a high-risk enterprise. The fruit handling systems for mangoes and bananas clearly show the role of simple technological innovations, such as those designed for heat dissipation, reduction of physical injury or disease control, in reducing risks (Concepcion, 1994).

Infestation by fruit flies (*Tephritidae: Diptera*), common in the tropics, is a major constraint to the production and export of tropical fruits. Current control procedures include cover sprays of insecticides, spot sprays of protein baits, orchard sanitation and fruit wrapping, all of which seek to prevent direct damage to fruits or are aimed at population suppression, mostly in individual plantations. These field control techniques enable fruit production of sufficient quality to meet the needs of domestic consumption as well as that for export to markets where fruit flies are not quarantine pests. (Vijaysegaran, 1994). It is worth noting that evidence suggests that in the EU quality standards have suggests that in the EU quality standards have been more systematically applied against imported products rather than to what is domestically produced or marketed.

Diversification of Production and Agricultural Policy Reform

South America has benefited from a number of land reform measures, as well as from a general agricultural and fiscal policy encouraging the production and marketing of fruits for export. The fruit sector of several countries in this area has expanded. Chile, for instance, has developed this sector since the introduction of structural reforms in the 1970s aimed at market liberalization. Fruit exports have increased four-to five-fold, making Chile a prime example of boom in fruit production and exports. The development of the country's temperate fruit industry has been quite impressive-pears, grapes, apples, and the rapid introduction of new products such as kiwifruit in 1987. A similar evolution occurred in Argentina when in the 1980s the government began encouraging producers of cotton, sugar and tobacco to diversify into fruit and vegetables (OECD, 1996).

Exports of Vegetables

Exports of vegetables in DCs increased at an average annual rate of 4.91% over the period 1986-1995, going fro-1995, going from 3.6 MMT in 1986 to 5.8 MMT in 1995 when this area originated some 36% of World total vegetable exports. However, data presented show a strong variability according to the country grouping, *i.e.* in South America the export growth rate peaked at 18.71%, while South Asia registered a 1.85% decline. Central America is the main exporting area (37% of total exports in 1995), followed by the Near East Asia (18%) and China (13%). One issue that affects very differently South American vis a vis Asian countries relates storage and transport to trade.

Technical Progress in Storage and Transport vs. Trade Liberalization

The growth in exports indicates that in Central and South America exporters, processors and producers are quite capable of adopting technology and finding new export markets. Poor infrastructure, lack of food standards and grading, and importer concerns about food safety are constraints which will have to be addressed in Asian countries. Although these problems can be considered as a limit to trade development, it is important to underline that trade liberalization has not developed at the same pace all over the world, and among the developing countries themselves the reasons are as much political as economic. A marked number of Asian DCs are not members of the Worldnumber of Asian DCs are not members of the World Trade Organization (WTO) and favour trade isolation. Export growth rates in most of the Asian country groupings are very low when compared with the corresponding level of production.

Vegetable Imports

On the import side the average growth rate of DCs has climbed to 5.15% (1986-95), going from 2 MMT in 1986 to

3.5 MMT in 1995 (World growth rate was 4.52%). In aggregate terms DCs represented in 1995 22% of World vegetable imports. The main importing areas were Near East Asia, with 37% of the DCs total import in 1995, and South Asia (33%).

The growth rate of vegetable imports is higher in South America with respect to Asian countries, where, however, import share is more consistent. The highly perishable nature of vegetables (as of all fruits) and the consequent need for immediate disposal calls for elaborate marketing arrangements in the absence of cold storage. South American countries have developed not only a good level of production but also an efficient storage and distribution system, and therefore can use raw products for the processing industry. In developing countries the vegetable processing industry has also important implications in relation try has also important implications in relation to employment.

Processing Industry and Employment

The processing of vegetables (as of fruits) helps to tackle the considerable potential loss arising from the perishable nature of the produce during times of seasonal glut in the context of low demand, poor infrastructures and weak transport. The reduction in tariff escalation helps to capitalize on off-season demand for high-value fruit in diverse developed country niche markets, enabling *horizontal* diversification away from traditional agricultural products. At the same time, it allows more value added in the form of further processing and packaging rather than the mere export of raw products, thus opening up in the process possibilities of *vertical* diversification by generating both rural and urban employment. Greater emphasis on a diversified and healthy diet has brought a substantial increase in the number and variety of fruit and vegetable products available to the consumer, with an ever increasing number of different types being processed and exported thanks to new heating (UHT, microwave, ohmic) and freezing (cryogenic) techniques combined with new packaging materials and technologies (aseptic, modified atmosphere packaging).

With the diminishing capacity of agriculture to provide employmentinishing capacity of agriculture to provide employment for ever-increasing surplus labour, the rural youth have been migrating to urban areas in search of a better life. The industrialization of rural areas is considered to be one of the most promising solutions to this problem. This is especially applicable to industries for the processing and preservation of fruits and vegetables (Khalil, 1993).

Export of Fresh Fruits

Fresh fruit exports by DCs (52% of World exports in 1995) grew at an average rate of 4.43% (World 4.32%), going from 14.2 MMT in 1986 to 18.1 MMT in 1995. Only North Africa showed a decline over this period (-2.61%). South America and Asian country groupings registered the most remarkable increases with a growth rate superior to 5%. In 1995 Central America covered more than one-third of fruit exports. With the share of South America (37%), these two country groupings originated some 70% of total exports by Developing Countries in 1995. Exports of temperate fruit and demand for tropical fruit products are two important issues affecting trade, with non-tariff barriers limiting potential imports from DCs.

Temperate Fruit Exports: Marketing Early or Late

Chile and Argentina are the largest exporters of temperate fruits (apple, pear, table grape and kiwi). To obtain favorable prices in export markets and to avoid tariff rates that otherwise could be very high, these countries take advantage of their seasonality, *i.e.* when their produce can be either early or late. Three types of methods are adopted to achieve this market advantage: cultivate early or late varieties (*e.g.* Chile harvests the early grape cvs. 'Perlette' and 'Flame' from early December essentially for the US market); adopt management techniques to bring harvest date forward *e.g.* plastic and hydrogen cynamide for grapes), adopt storage techniques to accelerate or slow the ripening process (*e.g.* controlled atmosphere, refrigeration and packaging film to slow and spraying ethylene to accelerate the ripening process).

During the last few years, sales by these countries have increased consistently. However, competition has also increased among them and with northern hemisphere countries. For instance, both Argentina and Chile are exporting the greater part of their pears to the EU, while the extension of the storage period for pears, currently nine months, has been an element of marked competition between producers in the northern and those in the southern hemisphere (OECD, 1996).

Demand for Tropical Fruits

Banana, Papgya, Guropical Fruits

Banana, papaya, guava and passion fruit deserve mention. The main banana suppliers are Costa Rica, Ecuador, Brazil, Honduras and India. Passion fruit producers are Ecuador, Brazil, Peru and Kenya. The major suppliers of mango are India, Peru and Ecuador, while guavas originate mainly from South Africa, Malaysia and India. The leading exporters of papaya are India, Peru and Brazil. For tropical fruits as a whole, the EU has been the fastest growing major import market during the decade 1986-96, with consumption more than doubling between 1984 and 1994. These fruits enjoy high income elasticities, *i.e.* a much higher consumption in proportion to the rise in income. A considerable amount of canned tropical fruit imported into the EU, however, is re-exported to other European markets, a substantial amount going to Central and Eastern Europe. Demand for tropical fruit products has, in recent years, risen on the international market, thus raising the stakes for certain developing country exporters.

Non-tariff Barriers

In general there are still non-tariff barriers that are limiting potential imports from DCs In this respect considerable importance is being attached to an international agreement on national sanitary standards and regulations. Maintenance of human and plant health are primary considerations in tropical fruit and vegetable-importing countries. The exporters have to satisfy consumer quality criteria as well as those of the relevant government agencies. The WTO's Sanitary and Phyto Sanitary (SPS) Agreement (separate from the Agreement on Agriculture) sets and encourages the adoption of international standards, while at the same time affirming the rights of countries to set their own health and safety standards if scientifically justifiable and not likely to result in an unreasonable barrier to trade.

Fresh Fruit Imports

Imports of fresh fruits in DCs grew at a yearly rate of 7.75%, doubling the amount traded (from 3.3 MMT in 1986 to 6.7 MMT in 1995). The share of DCs in the World aggregate went from 14% in 1985 to 16% in 1995. Remarkably, North Africa showed a 25.7% increase of imports, although the imports of this area account for only 1% (1995) of the total fresh fruit imports by DCs. Main importers of fresh fruit were the Nea6). Main importers of fresh fruit were the Near East Asia (34% of total 1995 imports) and East and South East Asia (29%) countries. It should also be noted that China is increasing its imports of fresh fruits (13% annual growth rate) and its share has grown from 3% in 1985 to 7% in 1995.

In Sub-Saharan Africa and South Asia the growth rate of imports is much lower because the capacity for self-production is higher. The relatively low growth rate in Asian countries is related to the high tariff duties on fruits.

The growth rates of imports by the various country groupings investigated are generally rather concentrated. This indicates that very differentiated production conditions are accompanied by similar trade situations, *i.e.* guarded opening to international trade. Until recently, most imports in South East Asia were channeled via Hong Kong, one of the area's few WTO member-countries.

The Underestimation of Chinese Trade Statistics

While China is a big importer of fresh fruit, official Chinese trade statistics vastly underestimate actual imports of fresh fruit as well as of many other commodities. Unrecorded shipments, rampant under-invoicing, misdeclaration of products, and other imponvoicing, misdeclaration of products, and other importation practices make Chinese trade statistics an inadequate indicator of real trade value and volume. Hong Kong's re-export statistics also reveal parts of China's import situation and usually are more accurate. Hong Kong re-exports are imports that transit the territory on their way to other countries, including China. As an indication of Chinese customs undercounting of fresh fruit imports, Hong Kong re-export statistics show much higher apple, grape, and orange trade volumes and values than the official Chinese data. In China and in many Asian countries most imported fresh fruit enters through "unofficial channels", because of phytosanitary restrictions and high tariffs. Under-invoicing and misdeclaration of product are two of the most common unofficial importation practices. As an example of misdeclaration, traders will grapes or plums-prohibited by a phytosanitary barrier-as apples, which can be legally imported (USDA, 1998).

Per Capita Vegetable Consumption

Per capita apparent consumption of vegetables showed a 0.92% increase over the period 1986-95, despite the growth rate decline in Africa and Central America. Apparent consumption went from 68.7 kg per capita in 1986 to 75.3 kg in 1995 on average. The highest consumption was registered in China The highest consumption was registered in China, where apparent per capita consumption jumped from 86 kg in 1986 to 146 kg in 1995 (a 5.38% growth rate). The lowest (and more stable) figure was registered by Sub-Saharan Africa (29 kg of vegetables consumed both in 1986 and in 1995).

New Consumer Trends in Vegetable Products

New trends in vegetable products for consumers in the DCs refers to people's standard of living. Generally speaking, fresh vegetables should lose market share to processed products. Many kinds of vegetables can be made into canned foods that cater to local tastes, cucumbers and peppers being but two examples. Easy to carry and convenient to serve, they can be stored for quite a long time and reduce losses incurred from seasonal supply surpluses of vegetables marketed at the same time of the year. At present the highest consumption has been registered by China, where three different kinds of vegetable should develop over the next few years-vegetable crisps, enriched vegetables and vegetable chips. Under vacuum-packaging at low temperature, radishes, melons and potatoes can be dehydrated and fried in edible oil into crisps. As they are processed in a low ambient temperature, which also reduced the loss of aromatic substances in their natural pigments, they can usually keep all their original flavors. They are regarded as a nall their original flavors. They are regarded as a new generation of food products upgraded to substitute the traditional fried and sponge foods (USDA, 1998).

On the other hand quite a number of scientists are doing research on so-called enriched vegetables, which contain many different nutrients. By means of conventional agricultural techniques, a certain number of trace elements can be dripped or sprayed onto the roots or leaves of vegetables to enhance them with the nutrients they lack. So far, Shanghai has developed a kind of vegetable called selenium cabbage.

Per Capita Consumption of Fresh Fruit

Per capita apparent consumption of fresh fruit has been relatively stable, with a growth rate of 0.38% (1986-95). Again, China has registered the highest growth rate (6.43%), while African and Near East Asian countries have shown a decrease. It should be noted, however, that in China consumption is comparatively lower (54 kg per capita in 1995) with respect to the other areas, *e.g.* in Near East Asian countries apparent per capita consumption is three times higher. Apparent consumption went from 107.96 kg per capita in 1986 to 109.35 kg in 1995 on average.

Chinese Consumer Trends

It must be assumed that the growth in per capita consumption in the DCs is the result of China's marked increase. According to trade sources, Chinese consumers buy most of their fresh fruit at street retail shops and wet markets, *i.e.* approximately 95% of imported fresh fruit purchases are made at these locations. US and European fresh fruit has won brand-name recognition by China's consumers, especially Red Delicious apples, Sunkist® oranges and Red Globe table grapes. Sunkist is one of the few overseas fresh fruit brands that consumers recognize. As an indication of its high level of local recognition, its packaging design has been imitated by a Chinese company with the intent to deceive local consumers. With the recent explosion in the number of supermarkets and hypermarkets, imported fresh fruit has begun to appear on retail shelves. While the amounts are still limited in comparison to those of the street stalls and wet markets, they could rise rapidly in the future (USDA, 1998).

Food Demand Projections

It should be noted in general that food demand in the DCs is projected to grow rapidly in the medium term due to a combination of factors-fast population growth, continuing rural migration to major urban centers and improved diets. Higher per capita income in these countred diets. Higher per capita income in these countries will provide impetus for a better and more varied diet, including to some extent fruits. Demand for fruit is generally very price elastic, thus final consumption is very much determined by the price, the quality and the disposable income of the consumer. (OECD, 1996, 1998).

Vegetable Exports and Imports

The total value of vegetable exports by DCs was US $3 billion in 1995 (28% of World aggregate), almost 50% of which originated in South America and 14% in China. The 1995 vegetable imports by DCs had a total value of US $1:4 billion (12% World aggregate), most of which originated from Asian countries (71%). The share of the East and South East Asian countries increased from 32% in 1985 to 42% in 1995, while the Near East in Asia registered an opposite trend, declining from 42% to 29% of total vegetable exports.

For vegetables the 1995 standardized trade balance, an index showing the propensity of a given area towards importing (0 ÷ -1) or exporting (0 ÷ +1) a certain commodity (or commodity group), was close to 1 in North Africa (0.96), China (0 was close to 1 in North Africa (0.96), China (0.91) and Central America (0.89). Asian countries, although registering a negative standardized trade balance, over the same period of showed a tendency towards a decreasing import propensity, *i.e.* Near East Asia went from -0.36 in 1985 to -0.08 in 1995.

Fresh Fruit Exports and Imports

More than one-third of World fresh fruit exports originated in DCs, although over the long term their share decreased from 43% in 1995 to 37% in 1995. South America is the main exporting area (33% of total DC exports in 1995) followed by Central America and the Caribbean (27%).

In 1995 DCs accounted for some 17% of World fresh fruit imports, an increased share with respect to the beginning of the observation period (14% in 1985). In 1995 the main importers were Asian countries, which accounted for more than 60% of the total imports, and South America (17%).

Central America showed in 1995 the highest standardized trade index (0.82) in fresh fruits, followed by African countries (0.64 in North Africa and 0.50 in Sub-Saharan Africa) and South America and 0.50 in Sub-Saharan Africa) and South America (0.37). China registered instead a negative 1995 index (-0.15), its import propensity also declining with respect to the previous decade when the trade index was at -0.59 (1985).

Conclusion and Prospects

Over the past decade increased trade volumes of fruit and vegetables, together with significant changes in trade flows, have been indicative of a general increase in consumption and a greater demand for imports, with a corresponding general rise in production. Linked to these aggregate trends has been a variety of shifts in demand, production and trade flows both within and between Developing and Developed Countries.

A limited number of DC country groupings dominates production and trade. There is competition between them and their relative shares of world production and trade shifted from time to time. The limited number of competitive country groupings can be explained in part by the agro-ecological characteristics of the areas under investigation and in part by the sum of factors comprising the cost of production, processing, marketing, availability of appropriate technologies and infrastructure, product standards and technical regulations. The constraints to increasing horticultural exports from India, for instance, include poor Post-harvest management, inadequate cold storage and prpostharvest management, inadequate cold storage and processing facilities, high import duties on packing materials, underdeveloped market infrastructure, inadequate market information services, low levels of productivity and fragmented holdings (Subrahmanyam, 1994).

No less important are entry obstacles to export markets resulting from the specialized nature of the export-related infrastructure, including quality and health standards, as well as established consumer preferences for specific products in particular markets.

During the period under investigation, trade volumes were stimulated by a general trend towards liberalization as well as an extension of trading blocs. The general expansion in the demand and trade of fruit and vegetables has been accompanied by a greater internationalization of trade due to a greater number of importing countries as well as supplier countries engaged in international trade, which in turn has led to an evolution of trade flows. In many cases trade that was formerly local or bilaterally oriented is now much more international, which has intensified the trading contacts between Developing and Developed countries. The greater participation of the former in this trade evolution, both in the role of supplier and consumer, is thus considered to be a major development of the past decade in world fruit and vegetable production and trade. Strong growth of exports has occurrend trade. Strong growth of exports has occurred in the traditional exporting area of South and Central America (in particular Chile and Argentina but also Brazil and Mexico). This trend is expected to continue in the medium term. Significant growth in exports has occurred and is expected to continue in Asia (especially China and India).

During the period under review, the Asian region has been the area of stronger growth, which, given generally rising per capita disposable incomes and the vast potential of the Chinese market, should continue as a major growth market.

Apart from China, the wave of development is the result of the fast pace at which the so-called tiger economies of South East Asia have grown during the last decade. With a population of about 600 million, excluding China and India, this part of the hemisphere is ideal for the fruit and vegetable business from the quantitative point of view. Since the early eighties the countries of the region have passed through an unprecedented boom with 15-20% annual growth rate, stable currency and low wages. Unfortunately some countries have run up a great deal of currency debts and the boom has been followed by a heavy downturn. This does not mean that the area is no longer of interest for worldwide trade but certainly in the near future trade will suffer the lack of purchasing power in the most important countries of the area. This is a market which rematries of the area. This is a market which remains of great interest for mass-consumer goods like fruit and vegetables.

This trend is expected to accelerate both in the short and in the medium term as a result of the Uruguay Round Agreement, which will provide (a) increased market accession, (b) reduce subsidized export and aggregate levels of support, and clearer and more detailed rights and obligations for (c) food safety and phytosanitary regulations. Although at this stage it is too early to estimate the URA's impact on fruit and vegetable developments, and actually this will vary according to the country and to the product concerned, it is evident that these three elements will stimulate the trade environment.

The further extension of trading blocs such as the North American Free Trade Agreement (Nafta) linking Canada, Mexico and the United States, various association agreements in Central and South America: the Central American Common Market, the Andean Pact, the Caribbean Community and Mercosur, the Association of Southeast Nations (Asean: Indonesia, Philippines, Singapore, Thailand, Brunei), and the Lome Convention between the European Union and the ACP countries are likely to provide additional opportunities for increased trade.

Future prospects of the fruit and vegetable exports of DCs will also largely depend on the growth of import demand, mostly in the developed countries. Dort demand, mostly in the developed countries. Developed countries are expected to diversify their food consumption patterns, including consumption of horticultural products. This heightened concern with health and nutrition and familiarity with more fruits and vegetables because of wider availability, increased travel and improved communications will lead to an increase in the ratio of imports to domestic products in total demand. Differences in per capita consumption of fruits and vegetables among developed countries with similar patterns and standards of living may also indicate a potential for increases in import-oriented consumption in countries where consumption is relatively low at present.

Look North to Canada as Promising Export Market

Trade in a foreign market doesn't always mean dealing with the Far East or Europe. A good place to start for U.S. exporters may be our northernmost neighbour—Canada. The country's proximity to the United States, and its common language and lifestyle, could make it a good place to gain valuable export experience and make sales.

The U.S.-Canada Free Trade Agreement (FTA) awakened many small-and medium-sized U.S. food companies to the fact that an important export market was right next door. Canada is the second largest market for U.S. agricultural exports and U.S. exporters are the dominant suppliers of agricultural exports to that country.

Of the $4.8 billion in U.S. agricultural exports to Canada in 1992,(*) 75 percent were in the high-value consumer foods category. In addition to fruits and vegetables and their products, the United States is the major source of other high-value consumer foods, such as beef, poultry, bakery products and pet food. Corn, soybean meal, vegetable oils, feeds and fodders and sweeteners are traditionally the major bulk and intermediate products exported to Canada.

Changing lifestyles in Canada are increasing the demand for U.S. agricultural products. With more women in the work force, there has been a reduction in the number of meals consumed at home. This has led to increased imports of U.S. consumer-ready products. Also, consumer awareness of health, nutrition and calorie-reduced foods, along with interest in international foods, has accelerated demand for U.S. packaged foods.

Fruit and Vegetables

Chile: Leader in Southern Hemisphere Fruit and Vegetable Exports. In the 1950s, the fruit and vegetable industry began a seedy export-oriented growth trend. Technological developments were the determining factor in launching this important production industry into the international arena.

Chilean fruit and vegetable exports date back to the early 20th century; however, the industry's steady growth did not begin until the early 1950s. The main factors that promoted this development were technological advantages that allowed shipping the products in vessels with cold storage chambers. The fruit and vegetable industry began in the late 1970s and early 1980s and it was due mainly to two factors:

- Chile 's natural conditions for producing fruit and vegetables that grow in mild climates.
- Changes in Chile 's economic policies, based on a strategy involving participation and opening up to foreign trade.

Advantages for Fruit and Vegetables

Chile enjoys a series of advantages for producing fruit and vegetables:

- Favorable soil and weather conditions
- Ideal phytosanitary conditions thanks to its geographic isolation originating in natural barriers (the Pacific Ocean, the Andes Mountains, the Atacama Desert and Antarctica).
- Counter seasonal production in contrast with the main consumer markets in the Northern Hemisphere.
- High standards among entrepreneurs and a qualified labour force, which translates into the availability of a skilled labour force.
- Low and rational use of agrochemicals. Economic policies that favour free enterprise. Diversified and staggered production.
- Constant introduction of new varieties, production technologies and preservation of fruit and vegetables, added to adequate shipment methods (sea, air or land) that ensure an uninterrupted year-round supply.

Evolution of Exports

Shows how fruit and vegetable exports have evolved, from US$168 million in 1980 to US$1679 million in 2003, *i.e.* an almost 9-fold growth in a period of 23 years.

Evolution of Fruits and Vegetables Exports

Year	*US $ millions FOB*	*Thousand of tons*
1980	168	261.0
1985	356	533.0
1990	716	1,034.6
1995	1,162	1,310.1
2000	1,386	1,506.4
2001	1,359	1,684.7
2002	1,534	1,830.5
2003	1,679	1,980.2

Source: Chilean Central Bank

In the mid 1960s, Chile ‘s share of all exports of grapes, kiwi fruit, apples and pears from the Southern Hemisphere amounted to only 3.5 percent. At present, Chile is the market leader in terms of volume, and it holds a 48 percent market share with regard to these species.

Chile is the main worldwide exporter of grapes and the second largest exporter of apples and pears.

In the Southern Hemisphere, Chile is the largest exporter of apples, peaches, berries and avocados, ranking second in pears and kiwi fruit.

Notably, 24 percent of the world’s grapes and 9.1 percent of the world’s peaches are produced in Chile. They are harvested when these fruits are out of season in the Northern Hemisphere, thus making Chile the main exporter for these products in the opposite season.

In the case of vegetables, specifically, average exports surpass US$31 million and the total volume is around 68,000 tons.

The main vegetables exported by Chile are onions, garlic, radicchios and asparagus.

A comparison of fruit and vegetable exports in 2003 with those made in 2002 shows that export amounts increased by 9.5 percent while export volumes grew by 8.2.

Fresh Fruits and Vegetables Exports

	Volume (Thousand of tons)				Value (MM US $ FOB)			
Species	2000	2001	2002	2003	2000	2001	2002	2003*
Grapes	596.2	545.3	654.9	730.0	662.5	580.4	645.6	710.0
Apples	387.7	542.7	548.2	585.0	202.2	237.2	275.1	305.0
Pears	124.8	128.8	119.4	122.5	64.4	60.1	63.0	68.0
Kiwi fruit	102.5	110.4	117.5	107.0	68.3	65.9	86.7	85.0
Plums	52.7	81.6	75.9	80.8	64.8	70.7	68.9	70.0
Nectarines	43.8	52.0	51.1	58.5	44.8	44.8	44.5	46.0
Peaches	30.8	37.7	37.6	44.4	25.9	35.3	32.9	35.0
Cherries	6.1	7.5	12.8	11.0	22.6	29.6	37.5	35.0
Avocados	52.0	52.5	78.1	80.0	73.8	79.6	97.4	110.0
Raspberries	3.6	4.1	4.3	4.5	21.1	21.3	23.2	25.0
Blueberries	4.0	4.4	6.4	6.4	12.7	13.9	18.1	24.9
Lemons	18.0	21.0	25.9	30.0	14.1	20.0	31.1	35.0
Onions	29.3	34.7	39.4	51.0	7.9	10.4	10.3	15.0
Garlic	10.3	8.9	10.7	11.0	9.1	8.5	11.7	10.1
Radiahios	2.3	2.3	2.5	2. 8	4.2	3.7	2.6	2.6
Asparagus	3.6	2.9	2.1	1.7	5.5	3.9	2.9	2.0
Other fruit	26.5	33.5	41.4	52.1	74.6	66.1	80.6	98.8
Other vegetables	12.2	16.4	2.3	1.5	7.1	7.9	1.7	1.3
TOTAL	1,506.4	1,684.7	1,830.5	1,980.2	1,385.6	1,359.3	1,533.8	1,678.7

Source: Chilean Central Bank/ *2003 Forecast

There are major changes in terms of the varieties of fruit grown and exported. For instance, Crimson Seedless is a new variety of table grape recently introduced that is expected to continue growing. Traditional groups such as Delicious apples have gradually been replaced by two-coloured varieties for which demand is currently rising, such as Gala, Royal Gala, Scarlett, Fuji and Braeburn.

Other Prominent Species

Species with considerable expansion rates in terms of export volumes include avocados, blueberries and lemons. The underlying reason is an increased surface area farmed in the last few years.

Main Species and Destination Market

At present, Chile makes direct export of more than 75 fruit and vegetable species to 70 markets. The main species exported are table grapes, apples, pears, kiwi fruit, pitted fruit and avocados. As shows, the most important species is table grapes since it accounts for almost 36.9 percent of the total volume of exports, followed by apples with 29.5 percent of the total volume.

Volume (Tons) - 2002/2003 Season						
Species	North America	Latin America	Europe	Far East	Middle East	Total
Grapes	416,217	80,583	129,473	65,650	6,334	698,257
Apples	95,348	202,808	203,945	30,429	65,349	597,879
Pears	19,604	32,113	56,088	716	4,698	113,219
Kiwi fruit	20,508	11,194	72,385	8,180	1,517	113,856
Plums	31,293	15,006	20,688	9,757	1,287	78,030
Nectarines	37,746	9,147	8,379	1,644	86	57,003
Peaches	29,615	15,782	1,107	175	0	46,679
Cherries	8,352	1,410	1,380	2,743	2 8	13,284
Avocados	89,297	123	2,687	8	0	92,115
Blueberries	4,877	7	372	216	0	5,474
Lemons	15,133	0	219	14,026	0	29,378
Other Fruit	90,707	8,899	18,094	9,137	778	54,070
Total 02/03	784,597	377,072	514,817	142,681	80,077	1,899,244
Total 01/02	671,426	440,881	437,592	128,145	74,670	1,752,714
Percent Variation	16.9%	-14.5%	17.6%	11.3%	7.2%	8.4%

Source: Chilean Exporters Association (ASOEX)

Markets for Fruit and Vegetables Export

In the 2002/2003 season, the main destination was North America (United States and Canada) with 41.3 percent of shipments sent there, followed by Europe and Latin America to which 27.1 percent and 19.9 percent, respectively, of the shipments were sent. As shows, the market share in other regions was Far East 7.5 percent and Middle East 4.2 percent.

Benefits of Free Trade Agreements for the Industry

The advantages of the Free Trade Agreement between Chile and the United States include the strengthening of the Generalized System of Preferences (GSP), a reduction in the high duties applied to avocados and asparagus, the establishment of a dispute settlement system, and a special committee on phytosanitary matters. In the Agreement between Chile and the European Union (FU), 85 percent of fresh fruit now enjoy tariff free status. This places Chile in a favorable position versus its competitors. Fruit and vegetables with high potential such as apples, pears, plums, avocados, raspberries, onions, asparagus and bell peppers can already enter the EU markets under full customs duty exemption.

Table grapes enjoy a benefit consisting of free quotas during two very important export periods: from January I to Icily 14 and from November 1 to December 31 of each year. Garlic was granted a 500-ton quota, with an annual increase of 25 tons.

Characteristics

Of all the different sub-sectors within the food industry, fresh fruit and vegetables have a chain of supply which displays unique characteristics;

- Fruit and vegetables are labour-intensive horticultural crops, mainly grown by small family enterprises. However, also banana companies and large USA fruit and vegetables enterprises are active in the market.
- In order to increase market power many of these smaller growers have organised themselves into sales organisations.
- Fruit and vegetables are traded as a fresh perishable products, makinlves into sales organisations.
- Fruit and vegetables are traded as a fresh perishable products, making the time factor and packaging critical.
- Nearly every single country in the world is involved in the trade resulting in a very large number of exporting and importing countries and a wide range of companies.
- The range of fruit and vegetable products is becoming increasingly wide.
- The range and origin of supply may shift from season to season. This unique situation of fruit in combination with vegetables, created in the retail product range makes it difficult for suppliers and retailers to establish agreements on long-term year-round supply and prices.
- A large number of retail outlets and formats exist, varying from large retail chains to small restaurants.

- Added value in the fruit chain is limited. This restricts the opportunity to create brands. The lack of clearly distinct consumer brands makes it easy for the buyer to shift from one supplier-trader-to another, and suppliers are therefore vulnerable. Buyers' loyalty to traders is based on service rather than on the branded products they sell.

Number and Size of Holding

The structure of the fruit and vegetable exploitation is not homogeneous. The complexity reflects the diversity of factors: evolution of consumption, production techniques, socio-economic interest, general politic climate, production policy, international competition and international trade agreements competition and international trade agreements.

Eastern European countries and many developing countries, still have a predominant traditional family-linked production unit. Traditionally fruit and vegetables were exclusively cultivated close to the cities, particularly for local consumption. Later transport over larger distances permitted specialisation and the de-localisation of production. Historical reasons explain the actual co-existence of different production systems:

- domestic production, particularly destined for auto-consumption, important in developing countries and in the Mediterranean area. Industrialised countries have garden plots and private orchards in areas where the climate is favourable for natural production.
- agricultural farms not specialised in fruit and vegetables but having a garden plot or orchard in order to obtain an additional income.
- professional enterprises, specialised in fruit and/or vegetables. A distinction is to be made between family holdings who use hired labour force for seasonal workers and professional non-family enterprises of larger size utilising one of more permanent employed persons and a large number seasonal manpower.

According to the results of structural analysis in EU in 1989-1990 approximately 1.8 million of farms produce fruit and vegetables. Surprising is the small average dimension: 1.3 ha. However, the analysis of professional exploit.3 ha. However, the analysis of professional exploitations shows that the average size is considerably larger. There are about 100.000 exploitation's specialised in commercial production of vegetables and 350.000 specialised in the commercial production of fruit. The average size is 4,2 ha for specialised exploitation's in vegetables and 7,9 ha for specialised exploitation's in fruit.

The specialised holdings achieve about 50% of the commercial production of the EU for vegetables and over 80% for fruit. The technological progress particularly in biotechnology and information did increase the operating scale and profitability of the exploitations. Particularly the integration of technological renewal in horticulture is important. The entrepreneur has to adapt his production methods and management in order to increase cost efficiency. The USA census 1992 reports 99.514 farms producing fruit and nuts, 29.605 vegetable (incl. melon) farms and 39.712 other specialised horticultural units. The average size of the USA farm is 21 ha for fruit and nuts and even 50 ha for vegetables and melons.

As compared to the EU, the USA 'family farm' dimension is significantly different. More-over, typical for both USA and EU, about 25 percent of the farms produce about 75 % of the production.

The requirement of capital per exploitation and per full-time employed person decreases per unit of the production. Thed person decreases per unit of the production. This evolution of production factors leads to the suggestion that the need for capital will increase and the need for labour will decrease. This is logically followed by a decrease of the number of exploitations. It is clear that the need for knowledge, comprehension and management capacities at the farm exploitations are still increasing.

Income Generation

Most fresh fruit and vegetables are produced by small and medium sized family farms. Once the seeds or trees have been planted, costs can be kept very low, particularly when labour is to a large extent supplied by the family itself. This makes the supply of fruit and vegetables comparatively less susceptible to price fluctuations; farmers will continue to supply the market, even when prices are only slightly higher than variable costs for several years. In the longer term, however, higher prices are needed to allow farmers to replant especially as planting new varieties is becoming increasingly important

Supply may vary greatly from season to season depending on the weather. According to economic theories a short supply would result in higher prices, which in turn counterbalances low volumes and therefore levels off the income effect. Fruit, however, from one destination or one kind may easily be substituted by fruit from other sources or kinds. Consumers may easily shift from one fruit to ainds. Consumers may easily shift from one fruit to another when prices for a certain fruit would be considered too high. Therefore producers and traders will rarely benefit from very high prices in years of shortages, and they will continue to experience the disadvantages of very low prices in years of abundance. This market situation with either low or very low prices results in structurally low margins for all parties. As a result of the price and marginal situation in the market, producers and exporters are constantly searching for possibilities to earn premium prices, and the role played by technology in these and other developments in the fruit chain is becoming increasingly important.

Supply

Worldwide, the 1996 production (FAO) reached 508 million tons for vegetables and 469 million tons for fruit. The productions dynamics is different according to continents, countries and types. Production is developed countries showed a strong expansion in the period 1950-60, but is stabilising. The yearly production growth in developing countries is more important than the demografic growth. This expansion matches the growing demand in Asia, Latic America and the Northern countries (out of season, tropical etc.)

Vegetable production increased by 56% over the last 15 years. Asia dominates with 60% of the world production. The EU comes next with 11%. The most importanroduction. The EU comes next with 11%. The most important producing countries are China, India, EU and USA. Over 100 types of vegetables are produced and consumed in the world. Only the types having good taste qualities and allowing for technical progress with respect to production and marketing are developed on a world scale. The breeding of vegetables, for example, is especially important for varieties cultivated in greenhouses or destined for further processing. This may result in standardisation of production.

Production Volume, 1000 MT

	AV 93-95	GR % 87-94
Citrus	87.933	3,48
Bananas	82.217	2,74
Tomatoes	80.664	2,94
Grapes	55.299	-1,67
Apples	48.920	2,68
TOTAL FRUIT(*)	448.588	1,57
TOTAL VEGETAB/CENTER>		
TOTAL VEGETABLES	489.242	3,21

*Excluding nuts

The development of the fruit production shows a similar trend to that of vegetables, having recorded substantial growth from just over 300 million tons in 1980 to in excess of 400 million tons in 1995, an increase of more than 30% over the last 15 years.

Growth rates varie largely. Grape production decreased by 18% during this 15-year period, whereas production of citrus, bananas and apples increased by 50%, 48% and 44% respectively.

The fall in grape production is quite an exception since most types of fruit have shown considerable growth. The largest growth rates were achieved by relatively smaller fruits, with the kiwi fruit having established itself as a major fruit following a rise in production of more than 3000% in the past 15 years. The more luxury fruit types have also experienced large growth: sour cherries and blueberries, for examples have shown growth rates of 102% and 98%. Tropical fruits have performed very well too, papayas growing by 81% and mangoes by 36%. Apart from mangoes though, all these fruits are still relatively minor components of the total fruit package. The above average production growth of the luxury and tropical fruits is a response to more sophisticated consumer demand in the developed ecosophisticated consumer demand in the developed economies that requires a wider variety of fruit throughout the year.

Production Value US $ 1.000.000

	AV 93-95	GR % 87-94
Citrus	22.976	3.7
Bananas	25.148	4.64
Tomatoes	31.902	5.13
Grapes	31.945	1.55
Apples	14.377	4.54
TOTAL FRUIT	134.599	1.77
TOTAL VEGETABLES	159.237	5.25

It is not only demand that has given momentum to fruit production, but policy has also stimulated fruit production, either directly or indirectly. Changes in the agricultural policy of many countries have led to farmers diversifying production and many have opted for cash crops sifying production and many have opted for cash crops such as fruit and vegetables. In North America, for example, the support for cereals and oilseeds was reduced in the mid 1980s, and this resulted in a subsequent increase in acreage under fruit trees in certain areas as well as diversification towards fruits of a higher added value such as berries. Chile is another example where structural reforms introduced in 1974 gave farmers greater security on their farmland property rights. Prior to 1974, annual crops tended to be grown because of the farmers' unwillingness to invest. But allowing farmers to own the land provided the incentive for them to invest in perennial crops and, as a consequence, fruit production boomed.

Supply Policy USA and EU

EU Market Management

More than in the other agricultural sectors the EU production of fruit and vegetables is guided by the market forces and influenced by tariff barriers. Therefore, for most products the surplus of production is minor.

There are however exceptions to this general rule. The EU withdrawals (below) are important and did become routine for a number of varieties of apples and peaches. Also in the case of citrus the withdrawals are relatively small but a growing part of the production is not suited for the fresh markets and is diverted (without support) to processed products.

The interdependence of producers of fruit and vegetable. The interdependence of producers of fruit and vegetables on the market influences the future evolution: if the internal market becomes oversupplied, a price drop will follow and lead to a progressive adjustment.

An external factor of increasing concern is the reform of the Common Agricultural Policy. A number of cereal producers are attracted to diversify their production to fruit and vegetables. They take advantage of the larger size of their enterprices and are supported by the EU for their traditional production. A similar situation is reported for vineyards, taken out of production by EU subsidies ('grubbing') and converted to production of apples.

The European withdrawal system was designed to cope with the cyclical surpluses unavoidable owing to the seasonal variations in yield due to weather conditions. Up until 1968 the withdrawal was entirely financed by the producers organisations. Modelled on the Dutch experience, the EU experimented with various forms of market management including :

- leaving the fruit on the tree and the vegetables in the field;
- setting increasingly demanding marketing requirements (*e.g.* minimum sizes);
- destroying part of the crop.

In practice the intervention by the EU was limited to setting the basic price and buying in price for the following produce only: eating apples, apricots, dessert pears, peaches, nectarines, table grapes, citrus, toert pears, peaches, nectarines, table grapes, citrus, tomatoes, cauliflower, aubergines, courgettes and cucumbers. A preventive withdrawal system has been in operation since 1979 for apples and pears as these can be stored. This mechanism was introduced as it became apparent that it was better to prevent a market collapse than to react once a crisis had occurred. In 1987 the EU begun to implement stabilising measures because of the fact that the cyclical events turned out to be a structural characteristic. Indeed, almost the entire crop of white peaches and red oranges ended up in withdrawal from the market. The withdrawal system was gradually loosing the initial aim of curing cyclical aspects. Consequently, the budget was rising to approximately, one million ECU on a yearly basis.

Moreover the public opinion became very dissatisfied with the size of the market withdrawals and particularly with the destruction of lots of products. The 1996 EU-reform of the fruit and vegetable market organisation is aimed at bringing a better equilibrium between supply and demand in the European Union. Budget saving are not the first goal. But the actual amount (2 billion ECU) would better be spent in order to satisfy producer and consumer.

USA Market Management

Most crops are subject to federal (USDA) or state quality standard. Both shipper and buyer are familiar with the USDA requirements shipper and buyer are familiar with the USDA requirements to be met for certain grades or sizes. If, on receipt of the product the buyer is satisfied that it is the requisite quality level, the product is accepted without any government inspection or involvement. The non-mandatory quality standards are simply there to facilitate trade, keeping direct costs down. One of the most important sources of institutional support for the US fruit and vegetable industry are mandated marketing programmes. Growers are free to choose from a menu of legal organisational options designing the programme that best meets their needs:

- Quality control improving the product sales and volume management provisions such as permitting only a certain portion of the crop to move into specified outlets (*e.g.* producer allotments);

- Market facilitating provisions such as production research, market research and development, market information;
- Market promotion and advertising.

All of the firms producing, and in some cases handling the commodity in a region encompassed by a mandated programme are required to pay an assessment levied on each unit sold. The funds are then pooled and administered according to the purpose agreed upon by the commodity group. The interest of the growers and/or handlers are represented by their elected grower and/or handler boards of directors.

Growers and/or handlers continue to sell their produc

Growers and/or handlers continue to sell their products independently as usual, merely contributing assessments on each unit sold. Promotional programmes are always conducted on a generic basis-promoting the product rather than any particular brand -.

Fruit and vegetable orders are active for: Florida citrus, limes and avocados, Idaho-Oregon onions and South Texas onions, all approved prior to 1991. It has been many years since new marketing orders with quantity control have been approved and it is viewed as highly unlikely that any will come into being in the future.

State marketing orders are required to fund production research through the corresponding state (Land Grant) universities. Research and promotion funds for fruit and vegetables collected under federal marketing order and check-off programmes are estimated to have totalled $ 45 million annually. The total farm value of the US fruit and vegetable industry was $ 23.000 million in 1994. Hence in relative terms, the industry investment in research and promotion through federally authorised programmes amounted only about 0.2 percent of the total cash receipts.

Importance of Producer Organisations

A new EU market management system was introduced in 1996. The institutional withdrawal prices are reduced gradually by 40 percent (30 percent for apples and pears) over a period of six years recent for apples and pears) over a period of six years. The withdrawal prices will be fixed for the entire EU as from 2002 on.

The EU 1997-reform is aimed at promoting the producer organisation

Key elements of the reform are :

- new responsibilities of the producer organisations;
- operational programmes;
- reform of the withdrawal system;
- interprofessional organisations;
- respect of environment.

The main objectives of producer organisations are:

- to market all the products produced by its members and to make sure that the production is planned and adapted to the consumes demand in the market, especially concerning quantity and quality; producer organisations receive EU funding on a 50/50 basis in order to realise operational programmes:
- to improve the quality of the products, to increase the commercial value of the products, to encourage the consumption, to develop the biological production, to encourage integrated production and to limit the withdrawal from the market;
- to take measures to encourage the producers to apply environment friendly production techniques, for the production as well as for the management of inputs;
- financial planning in order to apply the quality standards, the phytosanitary regulations and the maximum residue levels.

The creation of agricultural cooperatives in the USA is supported by law. Despite the appatural cooperatives in the USA is supported by law. Despite the apparent appeal of cooperatives, they are no longer a major part of the institutional setting in the fresh fruit and vegetable industry. Marketing cooperatives have been more important to the processing than the fresh market crop and particularly to the citrus sectors. In general the declining appeal of cooperatives is probably partly related to the increase in average grower size, as larger growers are able to market their crops independently.

Supply of Some Important Crops

Citrus

World citrus production totalled 87,9 million tons in 1994, 23% of all fruit produced. It can be seen that the gradual increase in production occurred mainly after 1985. For grapefruits, most of the growth occurred during the nineties, a rise that can be explained by the development of new, less bitter varieties that are better adapted to the taste of consumers.

Differences exist between oranges grown for processing or for direct consumption. Fruit sold on the local fresh fruit market often yields a better price than fruit sold to the processors, fresh fruit demand needs to be satisfied before oranges are supplied to the processing industry.

Encouraged by good producer prices, many orange trees were planted in Brazil between 1988 and 1993 but lower prices in recent years have discouraged the planting of new grapes and may curtail priscouraged the planting of new grapes and may curtail production growth in the future. In the USA citrus production is moving south within the state of Florida to reduce the risk of frost.

In Mexico and Italy most of production is consumed domestically. In Italy and the USA 8% of citrus production is used for export purposes, whereas Spain is the only country which mainly produces for the export market with more than half of production exported.

Citrus Production, 1000 MT

	Production AV 93-95	GR % 87-94	
Africa	7.423	1,80	
Latin America	30.780	3,66	
Asia	22.137	7,33	
Industrialised Countries	-	-	
EU	9.161	1,25	
NA	13.828	3,55	
TOTAL	87.933	3,48 </T3< CENTER>	3,48

Grapes

The fall in grapes production over the past 15 years is evident. With more than 70% of production used in wine making, the largest part of the production fall is attributable to the fall in wine production. This is clearly illustrated by the fact that grape production declined especially in Italy, Spain and France, the world's major and traditional wine producing country. Europe is responsible for 72% of world wine production, 52% of grape production and 58% of vine area. In the USA, Turkey and Southern Hemisphere countries a relatively high share of grape production is used for raisin production and fresh use.

In contrast with total grape production, the production of table grapes has remained rather stable during the nineties and is estimated at ten million tons. Production of table grapes is concentrated in four countries with Turkey, Italy, Chile and the USA (California) having a global market share of more than 60%.

Grape Production, 1000 MT

	Production AV 93-95	GR % 87-94	
Africa	1.290	0,14	
Latin America	5.414Latin America	5.414	-0,33
Asia	9.802	2,96	
Industrialised Countries	31.877	-2,06	
EU	23.506	-2,97	
NA	5.436	0,96	
Economies in Transition	6.916	-4,73	
TOTAL	55.299	-1,67	

Bananas and Plantains

World banana production increased from 43 million tons in 1978 to 54 million tons in 1994, an average increase of 3,12% per year. Ranking second and fifth among the world's major fruits, bananas and plantains are important crops. The plantain or cooking banana is a variety of banana which is part of the daily diet in tropical areas. With less than 1% of total production exported, the product plays an insignificant role on the world market.

Because of the banana's dual character, being produced both as a staple food and for export racter, being produced both as a staple food and for export purposes the production structure is very diverse, often very small scale and not on a commercial basis because a banana tree can be grown in every suitable backyard. Banana production for commercial purposes varies from smallholder production with farms often smaller than five hectares to production on plantations of several thousand hectares. The scale of banana production is to a large extent determined by the natural conditions.

In Central America, where most of the dollar bananas originate, production is fairly export oriented. Independent producers exist, but a considerable part of production is produced on plantations owned or controlled by one of the large multinational banana companies, the best known of which are United Fruit, Dole and Del Monte. The large-scale plantations, sometimes as large as 5.000 hectares, are characterised by efficient organisation, substantial know-how, fast implementation of innovations, high yields, high consistent quality and a relatively high capital input. Independent producers tend to dominate production in EU and ACP countries, varying from small scale production to large plantations. Most Caribbean producers have plantations of 5 to 20 hectares, though in some countries larger plantations exist, up to sizes of 2.000 hectares.

Exports from these countries are mainly organised through independent European importers (Fyffes, Pomona) Independent European importers (Fyffes, Pomona).

Banana Production, 1000 MT

	Production AV 93/95	GR % 87-94
Africa	6.581	1,77
Latin America	22.870	3,49
Asia	23.119	3,39
Oceanic	687	3,03
Economies in Transition	827	-1,86
TOTAL	54.083	3,12

Apples

The average annual world growth is estimated at 2,68 percent, particularly because of the production increase in

Main China. The world apple production for the year 2000 is estimated (Washington State University) at 62 million ton. Red Delicious is world wide the most important variety, followed by Golden Delicious. Fuji has a remarkable recent increase: 80 % of the total acreage of this variety, namely ncrease: 80 % of the total acreage of this variety, namely 370.000 ha, is planted in China. China had only a small production in 1987 and is now the world largest producer.

It is expected that the acreage of red varieties will remain stable in the year 2000 and so does the acreage of Golden Delicious. The acreage of bicolour varieties will likely increase. Red Delicious remains the number one variety for the USA. The share of Granny Smith, Empire, Fuji and Gala is increasing. Plantings of Pink Lady are started.

Production in EU the still increasing but there has been a shift from green and yellow varieties to red and bicolored varieties. Jonagold, Gala, Breaburn and Elstar are trend-setters. Standard Golden Delicious, Red Delicious, Granny Smith, Cox's and James Grieve are decreasing.

It seems that some regions are looking for typical varieties such as Jonagold in Belgium, Elstar in Holland, Golden Delicious in Limousin and Pink Lady in South France. There has been a switch to fruit growing by cereal farmers and wine grape growers as a result of lower guarantee prices for cereals and wine. Large scale intervention has occurred in the EU in bumper crop years (Sec 2.1.1.). Germany and Switzerland are looking for new resistant varieties particularly on farms with direct selling to consumers.

Eastern Europe is looking for varieties, suited for the Western European markets, such as Jonagold, Elstar and Pinova. But thearkets, such as Jonagold, Elstar and Pinova. But the majority of the orchards are out of date. The renewal is relative limited because of lack of capital, uncertainty on land-ownership, *e.g.* in the former Eastern Germany. In the Southern Hemisphere, Argentina used to be the most important producer. The SH has plenty of possibilities but the capital is lacking. Today Chile is the most import apple producer in SH with a relative decrease of green varieties and an increase in Gala,

Braeburn, Fuji and Scarlett. Some parts of Brazil are climatologically well suited for Gala.

Granny Smith is the main variety for South Africa, occupying 50 % of the plantations: the share of Gala is increasing. South Africa has the possibility to increase the production volume by 25%.

The production in New Zealand is stabilising after many years of increased renewal of older plantations has taken place. The Apple and Pear marketing Board has an important role in variety selection and quality policy; Pacific Rose is considered promising.

Apple Production, 1000 MT

	Production AV 93-95	GR % 87-94		
Africa	851	15,97		
Latin America	3.213<>	Latin America	3.213	3,52
Asia	18.815	9,69		
Industrialised Countries	17.833	0,47		
EU	9.563	-0,62		
NA	5.498	+2,42		
Economies in Transition	8.209	-3,74		
TOTAL	48.920	2,68		

Tomatoes

Over 80 million tons of tomatoes, 19% of the fresh vegetable production in the world, are produced every year. The world production of tomatoes only was 35 million tons in 1970. This increase resulted in part from the increase of area but for a more important part from technological innovations and the development of productions in glasshouses. Many countries are producers of year-round tomatoes and not only in the summer-months, using new to varieties adapted artificial heating and high technology.

Tomatoes are produced and sold in all zones of the world. Theatoes are produced and sold in all zones of the world. The EU is the most important producer with 13 million tons of tomatoes or 18% of the world production. The Mediterranean areas of the EU (Italy, Spain, Greece, Portugal, South-France are the most important producing areas. However an important part of the crop is produced by northern European countries (Holland, Belgium).

The USA is the next important producing country. Tomatoes are the USA third important product after citrus and grapes. California and Florida are the two important producing regions.

The Turkish production however has an exceptional growth rate going from 1,2 million tons in 1961 to 6 million tons in 1994. This is particularly due to the development of irrigation. The total irrigated area in Turkey represents 20% and an additional increase is to be expected.

Approximately 30% of the world production is destined to further transformation.

Tomato Production, 1000 MT

	Production volume	GR % 87-94		
Africa	8.732	2,53		
Latin America	7.584	2,47		
Asia	29.547	Asia	29.547	6,69
Industrialised Countries	26.914	2,7		
EU	12.792	1,43		
NA	12.021	4,46		
Economies in Transition	7.882	-4,62		
TOTAL	80.664	2,95		

Trade

Overview

Within the chain of supply, there are several trade steps between the fruit grower and the end consumer. A distinction is made between:

- Collection wholesaling (collecting the fruit from the grower).
- International trade
- Distribution wholesaling (distributing the fruit to the retail outlet).

A number of traders combine two or even three of these functions. Almost all vegetables and most of fruit in the world are consumed as a fresh product in the domestic market where domestic traders are active. Indeed, some of the largest fruit and vegetable growing nations produce almost exclusively for the domestic market. However, over 30 % of frumost exclusively for the domestic market. However, over 30 % of fruit is processed mostly into juice and wine, and these are often traded internationally as well (juice is traded predominantly in a pre-processed form as concentrate).

While fresh fruit is consumed all over the world no type of fruit can be grown everywhere. Since consumers demand a wide choice, trade must take place between producing and consuming regions, making the trade function particularly important. Nearly 9 % of fruit production is traded internationally as a fresh product. The proportion of fresh fruit which is traded increased from 6,9 % in 1980 to 8,9 in 1995. While this is low compared with other agricultural commodities such as wheat, it is high compared to vegetables which are predominantly grown in consuming areas.

Because fruit can be consumed both fresh and in processed forms, the volume available for international trade depends on a combination of factors including annual production, levels of processing and domestic consumption. Competition between these market segments, however, is relatively low as each segment consumes produce of a different quality.

Export Volume, 1000 MT

	Export AV 93-95	GR % 87-94		
Bananas	12.465		12.465	7,29
Citrus	8.913	2,14		
Apples	4.844	4,67		
Tomatoes	3.171	4,44		
Grapes	1.968	4,70		
TOTAL VEGETABLES	15.343	5,98		
TOTAL FRUIT	36.471	5,42		

With exception of the SH, most large export forces are close to the import markets. A distinction can be made :

- Mediteranean Countries (South EU, Turkey, Israel, Maroc) exporting to the Northern European Markets. The EU imports (1996) 45% of the world trade of fruit and 44% of the world trade of vegetables. The EU is also the largest world exporter of fruit (31%) and vegetables (31%);

- Countries of Central America (Mexico, Equator, Costa Rica), exporting mainly to NA (USA and Canada) markets. In this area, the banana export occupies a major role. The USA ranks second as largest world importer of fruit (ranks second as largest world importer of fruit (19,3%) and vegetables (19,8%);
- South Asian countries (China, Thailand) mainly selling to Japan and the new emerging economies.
- SH countries, very specialised during the last 19 years. They take advantage of the off-season effect on the North European and North American Markt and are characterised by large commercial organisations.

It is clear that the overall technical progress in transport, handling and storing offered large opportunities for importing countries to diversifie their sources of import

Export Value, US $ 100.000

	Value AV 93-95	GR % 87-94	
Citrus	4.479	-	
Bananas	4.010	10,22	
Apples	2.521	8,23	
Tomatoes	2.516	7,77	
Grapes	1.824	8,07	
TOTAL VEGETABLES	10.536 <>	10.536	10,22
TOTAL FRUIT	19.287	8,20	

Trade Policy in USA and EU

World horticultural trade is very much affected by all sorts of regulations; non-tariff barriers are complemented by a wide range of tariff barriers in the form of export taxes and import levies. Phytosanitary restrictions can be legitimately applied to ensure food safety and to prevent the spread of diseases, but they can also be implemented for the sole purpose of acting as a barrier to trade. Most trade regulations are established to protect domestic producers from foreign competition.

However, there is a marked general trend towards liberalisation of world trade. Numerous multi-lateral and bi-lateral trade agreements have been concluded over the years. The NAFTA and MERCOSUR agreements are only two examples which have boosted fruit and vegetable trade on the American continents. Within the framework of GATT/WTO there is also a continuous move towards a further liberalisation of world trade. The minimum access requirement, tariffication and reduction in export subsidies that was agreed upon in the latest GATT agreement, will boost fruit trade. A reduction in subsidies will disadvantage the least efficient growers. Production costs inntage the least efficient growers. Production costs in the primary sector will become the focus of competition resulting in a more significant role for global supply.

Despite these new incentives to liberalise world trade, there still is room for countries to protect their own markets and domestic producers from world market fluctuations. Even the GATT/WTO agreements have included some provisions that enable the participating countries to protect their markets. Even though all countries visibly protect their domestic produce from foreign competition, the EU is the most accused of being too protective and disruptive to the world market. The banana import regulation is the major instance of this and is treated more extensively in the sections dealing exclusively with bananas.

It is not only for bananas that the EU import regime is rather restrictive. At the beginning of 1996 the reference price system was replaced by the entry price system for a number of products. The EU replaced its fruit and vegetable reference price system by the entry price system to meet the GATT agreement which demands tariffication of import restrictions. The entry price system is additional to a universal import duty of 8%. With the entry price system, applied to 15 fruits and vegetables, the EU sets a daily minimum import price, depending on the prices of EU produced fruit. On products entering the EU below this minimum price a progressive import tarow this minimum price a progressive import tariff is levied to equalise the import price with the minimum price. On products entering the EU at 8% or more (12% for apples and pears) below the entry price, a fixed duty is added. The entry prices are consistently set at levels well above the production costs in the EU. The entry prices vary per product and per season.

In line with the GATT/WTO treaty, liberalisation of trade is expected to continue as part of the new WTO round after the year 2000. This will result in new market opportunities, in particular for those countries with a low cost price. A reduction of subsidies will penalise the least efficient growers and different national fruit chains will witness increasing competition on production costs in the primary sector. Global supply will acquire greater significance.

Trade of Some Important Crops

Citrus

Citrus export: In 1995, 9,7% of world citrus production was exported as a fresh product, a volume of 8,9 million tons. Some 8% of oranges was exported whereas for tangerines this was 13% for lemons and for limes 14% and for the combination of grapefruit and pomelo as much as 22%. Oranges account for half of citrus exports.

The export of tangerines in particular is very concentrated with Spain alone supplying more than half of the volume of trade. Most Spanish citrus exporte than half of the volume of trade. Most Spanish citrus exports are destined for other European countries. Spain has been able to increase exports substantially during the first half of the nineties. EU membership has enabled Greece to boost its orange exports. The decrease in Israeli exports after 1990 is said to be the result of the country's privatisation of citrus exports in 1991, the result of a hasty decision for which that the sector was poorly prepared.

Citrus Export Volume, 1.000 MT

	Export AV 93-95	GR % 87-94	
Africa	614	-3,24	
Latin America	862	-1,12	
Asia	967	3,69	
Industrialised Countries	6.401	2,93	
EU	4.350	4,15	
NA	1.201	3,89	
TOTAL	8.913	2,14	2,14

Citrus Import: The substantial fall in orange imports in Germany and France since the late 1980s has, however, been largely compensated for by imports of other citrus fruits. The import of tangerines in particular increased but total citrus imports remained fairly stable, only France showing a slight decline since about 1990. The shift away from oranges may be induced by the demand for convenience : tangerines are more easy to peel than oranges for example. There may also have been a shift away from fresh orange consumption towards consumption of orange juice, particularly the case in Germany.

Export prices from Greece are at the lower end of the market, whereas American oranges receive premium prices.

The growth in the world citrus trade is expected to continue in the future despite the stabilisation of imports in the developed markets. This potential for growth is due to the development and marketing of new citrus varieties which are better adjusted to consumer requirements for taste, size colour and quality. The so-called easy peelers seem to have especial potential which clearly is a result of the search for convenience.

Citrus Import Volume, 1000 MT

	Import AV 93-95	GR % 87-94< AV>	GR % 87-94
Africa	50	13,69	
Latin America	519	13,0	
Asia	865	6,09	
Industrialised countries	5.880	0,77	
EU	4.523	0,29	
NA	579	3,58	
TOTAL	8.171	2,10	

Grapes

Grape Export: Approximately 20% of world table grape production is exported. This is very high compared to the average for all fruits of 9%. World trade in grapes, most of which are table grapes, has shown a considerable increase during the last fifteen years, Chile alone accounted for almost half of the export growth by increasing its exports from 49.600 tons (1980) to 442.820 tons. Not only Chile, but also South Africa increased exports substantially, especially after sanctions were lift in the early nineties. Specially after sanctions were lift in the early nineties.

With three countries, the EU is well represented among the major grape exporters. Most of the exports from Italy, Spain and Greece (84%, 94% and 80%, respectively in 1995) are directed toward other European countries, with Germany the major destination. The dramatic decline of grape exports from Romania and Bulgaria, two countries that were among the major grape exporters in the early eighties should be noted. Their exports are now almost down to zero as a result of the economic problems in Eastern Europe.

Grape Export Volume 1000 MT

	Export AV 93-95	GR % 87-94
Africa	2	36,78
Latin America	520	6,87
Asia	118	2,75
Industrialised countries	1.291	5,13
EU	920	3,72
NA	258	9,41
<8	9,41	
Economies in Transition	37	-12,2
TOTAL	1.968	4,7

Grape Import: The northern hemisphere in all absorbs more than 98% of world exports and is a major destination for southern hemisphere exports. European Union imports, approximately half of world imports, were mainly originating from other EU countries. Imports from non-EU countries also increased due to a rise in demand. The former Czechoslovakia used to be an important grape importer, but due to economic problems its position has gradually weakened since the late 1980's. The USA holds an important position both as exporter and as importer, though it is a net importer. US exports mainly Californian grapes during the second half of the year and imports are concentrated between January and May.

The growth in international trade for grapes was encouraged by the development of new varieties. The Thompson Seedless was the first seedless variety to be marketed on a large scale.

The major importers, the USA and the EU, shield their domestic producers by imposing protective measures during their own production season. The EU for instance applies its entry price system to table grape for instance applies its entry price system to table grapes from 21 July to the end of November. Grapes imported during the EU's off-season are virtually unaffected. The USA has a similar methodology: table grape imports are subject to two marketing orders, specifying certain quality requirements, between 20 April and 15 August, in addition to the application of phytosanitary barriers.

Grape Import, 1000 MT

	Import AV 93-95	GR % 87-94	
Africa	3	13,55	
Latin America	76	-	
Asia	188	5,82	
Industrialised countries	1.468	2,73	
EU	929	2,45	
NA	480	3,49	
Economies in Transition	167	10,66	
TOTAL	1.922 <198>TOTAL	1.922	4,05

Bananas

Banana Export: With exports in 1995 over 13 million tonnes, almost 25% of production, bananas account for approximately one third of the world's total fruit exports. Most bananas are exported from Central America but the Philippines is also a major exporter. The largest exporting nations have shown the largest growth during the past fifteen years, suggesting that world banana trade has become more concentrated.

Banana Export Volume 1000 MT

	Export AV 93-95	GR % 87-94
Africa	346	9,22
Latin America	9.079	5,98
Asia	1.369	4,48
Industrialised countries	1.432	27,4
EU	1.043	44,12
NA	387/I>	
NA	387	11,85
TOTAL	12.283	7,29

Banana Import: On a world scale banana imports increased by more than 3,5 million tonnes between 1990 and 1995 despite the liberalisation in the EU's banana import regime that was implemented in 1993. With 37% of world imports, it is no surprise that the EU banana regime has given rise to so much debate.

Germany is the second largest importer after the USA, although the fall in imports since 1992 is probably due to the EU banana regime which resulted in much higher consumer prices.

Recent developments in world banana trade indicate strong growth in imports by Eastern European countries. This is mainly the result of the banana's image of a 'luxury' fruit while still being very affordable When the economies in these countries gather more momentum, there is likely to be a slow-down as demand for other fruits increases.

China is still a relatively unknown factor as banana importer but if it continues its present trend with imports increasing more than tenfold between 1990 and 1995, from 14.000 tonnes to 160.000 tonnes.

World banana trade is likely to increase further in the future, though there are some hints that the EU market is becoming s, though there are some hints that the EU market is becoming saturated, not only because of the restrictions on imports. In developed economies bananas may have to cede market share to other newer exotic fruits.

Banana Import, 1.000 MT

	Import AV 93-95	GR 87-94
Africa	11	23,94
Latin America	477	11,89
Asia	885	15,86
Industrialised Countries	9.733	5,55
EU	4.599	8,86
NA	4.013	3,27
Economies in Transition	1.558	39,63
TOTAL	12.705	8,10

Apples

Apple Export: The total word trade is estimated at 4,8 million tons yearly. About 1.2 million tons originates from the SH. The apple exporters not only look for increase in the existing markets but as for new markets, such as Eastern Europe and South East Asia.

It is expected that the total apple exports from the SH countries will reach 1.6 million ton by the year 2000. The SH export of Granny Smith is the most important with approximately 400.000 tons. The SH exports of Golden Delicious, Braeburn and Fuji are estimated at respectively 100.000 tons, 175.000 ton and 110.000 tons.

Particularly the USA are successful pioneers in looking for new markets in South East Asia: but Chile, New Zealand, Australia and Canada are also very active. The import of apples in Hong-Kong, Taiwan, Singapore, Thailand, Malaysia and the Philippines is increasing year by year.

The USA export to these countries is estimated at 300.000 tons in 1995 as compared to 230.000 tons in 1994. The European Union is not so active on the East Asian markets.

Apple Export Volume, x 1000 MT

	Export AV 93-95	GR 87-94	
Africa	0,3	35,08	
Latin America	585Latin America	585	1,76
Asia	459	9,92	
Industrialised Countries	3.192	6,17	
EU	1.992	4,71	
NA	707	12,81	
Economies in Transition	608	-1,30	
TOTAL	4.844	4,67	

Apple Import: The apple import in the European Union increases year by year. The imports exceed the exports: in 1994 the EU imported 566.000 tons as compared to the export of 403.000 tons. The imports in 1995 were 693.000 ton, consisting of 590.000 ton imported from the SH (South Africa, New Zealand, Chile and Argentina.

Many countries are active on the Eastern European markets: particularly the USA and New Zealand discovered this market. A large quantity of class II apples is exported to Eastern European countries. It's is exported to Eastern European countries.

China is considered to be a potential exporter of apples. The increasing production is not only for domestic consumption but China may want to become an important supplier for South East Asia (Fuji). Of course, as the Chinese economy grows, this country is considered to be a niche market for specific varieties as well.

Apple Import Volume, x 1000 MT

	Import AV 93-95	GR % 87-94	
Africa	51	14,15	
Latin America	389	13,64	
Asia	779	8,67	
Industrialised countries	2.653	1,16	
EU	2.359	1,43	
NA	229	-1,17	
Economies in Transition	1.085	9,5	
TOTAL	4.963TOTAL	4.963	4,53

Tomatoes

Tomato Export: Only 4% of the world production of tomatoes is exported fresh.

With approximately 700.000 tons, the Netherlands is the leading exporting country of the world. Ten percent of the export is during the winter months, coming from Canaries. Spain is the number two export country together with the Canaries providing 45% of the northern European market during the winter months.

There has been a clear increase of Spain tomato exports during the recent years.

Moroc takes advantage of the free access to the European Union. Mexico is the number three export country in the world and is almost exclusively the supplier of the USA (98% of the import of the USA). The Mexican imports cover approximately 20% of the USA fresh tomato consumption. The export of the USA comprises 225 thousand tons going mainly to Canada.

Tomato Export Volume, 1.000 MT

	Export AV 93-95	GR % 87-94		
Africa	182	6,63		
Latin America	580	<>	580	0,95
Asia	370	2,52		
Industrialised countries	1.957	7,6		
EU	1.769	7,12		
NA	173	13,32		
Economies in Transition	83	-11,4		
TOTAL	3.171	4,44		

Tomato Import: Germany is the number one import country of tomatoes in the world. Approximately 55% of the German import originates from the Netherlands. Next is Italy and the import originating from Spain is increasing rapidly.

The number two import country in the world is the USA and the trade agreement between Mexico, USA and Canada increases trade. Surprisingly Poland and also Russia have a very strong yearly growth of import of tomatoes. The disruption of their economy and the opening of their markets gave an opportunity to Benelux exporters to increase their yearly export.

Tomato Import, 1.000 MT

	Import AV 93-95	GR 87-94
Africa	11	15,67
Latin America	56	9,7
Asia	402	0,57
Industrialised countries	2.295	3,5
EU	1.595	4,03
NA	630	1,98
Economies in Transition	320	14,05
TOTAL	3.083	3,95

3

Demand for Fruit and Vegetables

Structure

The average world consumption of fruit, excluding nuts, is 79,4 kilo per person per year and 84,9 kilo per person per year for vegetables. Important disparities exist between countries and continents. So one disparities exist between countries and continents. So one quarter of the world population consumes more than 100 kilo per person per year of fruit.

Population and Consumption of Fruit and Vegetables, 1994

	Population x 1.000.000	GR % 87-94	Vegetable consumption kg/p/y	Fruit consumption kg/p/y		
Africa	660	2,78	40,1	69,1		
Latin America	469	1,79	42,7	74,3		
Asia	3.188	1,74	92,5	54,4		
Oceanic	7	2,15	69,3	204,7		
Industrialised Countries	869	0,71		0,71	116,6	132,4
EU	370	0,37	130,7	171,1		
NA	294	1,03	118,3	125,7		
Economies in Transition	414	0,37	86,5	52,4		
TOTAL WORLD	5.607	1,59	84,9	79,4		

These different levels depend on two factors:

- the climatic and agronomic possibilities.
- the buying power and commercial infrastructure.

Three types of consumption could be distinguished:

- Countries with high consumption (over 200 kilo per person per year).

Those countries make advantage of the natural conditions favourable for the culture of fruit and vegetables and are mostly rich countries. From the end of 1950 the increase in consumption is linked with increase in per capita income. The consumer choice is diversifying but co-capita income. The consumer choice is diversifying but consumption is not increasing in volume any more in the last 10 years. In this group we find the countries of the Mediterranean area, the industrialised countries of NA and EU and the Middle East countries taking advantage of oil dollars. Also Arabian islands, banana producers and some Asian countries (Singapore, Korea) have similar trends.

- Countries with average consumption (between 100 and 200 kilo a person a year).

In this group we find countries with weak revenue, traditional consumers of fruit and vegetables or countries with higher incomes but not having favourable natural growing conditions. Among those the former USSR and the Eastern European countries having financial problems by the liberalisation of their economies as well as some African countries (Rwanda, Congo).

- Low consumption countries.

These are the countries with low buying power, situated in Asia and Africa. For a certain number of them the level of consumption is underestimated because of non-known level of auto-consumption and buying out of traditional commercial structures. The countries with the lowest consumption are the Sahel countries (Africa, Mauritania, Somalia, Australia and Yemen) and the cold and hilly areas (Mongolia and Afghanistan).

Stagnation and Individualisation of Consumption in EU and USA

Demographic evolution in Western Europe and the increasing age structure of the population leads to the conclusion that a relative saturation is occurring in terms of quantity of demand.

The smaller size of the households and the time pressure leads to the demand of convenience food. The demand for new technologies and Post-harvest handling techniques is high.

Elements are:

- changing menu structure
- individualisation of consumption patterns
- smaller households

One factor is the demand for adapted small packages. The suspicion of the consumers for new processes and raw materials has increased demand for information on production processes and raw materials.

Three major elements are to be mentioned:

- differences between countries
- general tendency to increase consumption
- general tendency to diversify consumption

The differences of consumptions between countries the members of the European Union are considerable. The relation between the country with the highest consumption of vegetables and the lowest consumption is one to five for vegetables and one to six for fruit. In general the countries with high production tend also to have a high consumption. This is the case for the Central European countries.

In 1994 the EU consumption of vegetables, both fresh and processed was 130,7 kg per person. About 26% of global consuesh and processed was 130,7 kg per person. About 26% of global consumption is processed. The deep frozen market is most dynamic in vegetables: this segment is still highly expansive.

Trends in fruit consumption are difficult to determine: for both EU fresh and canned fruit the last estimations are 171 kg per person. The consumption of processed fruit, more especially in the form of juice, is important for a number of fruits. As for the total, the consumption of fresh and processed fruit increases by approximately 1.2% per person during the last years. It is to be expected that this increase is slowing down. A yearly increase is expected by 0.9% at the end of this decennium.

Marketing Services

Fruit and vegetable distribution proceeds through three different channels:

- a "short" channel of direct sales and local markets;

- "traditional" distribution involving the physical presence of the products on wholesale market (*e.g.* Rungis or Brussels);
- "integrated" channels which are becoming more and more important.

Direct producer sales to retailers or caterers and sales on local markets ("short" channels) have developed above all wherever growers are located close to towns or tourist regions, giving them as a rule a good rate of return for their crops.

In the EU, direct household purchases from the farm average about 5 per c, direct household purchases from the farm average about 5 per cent of total purchases, apart from a few countries such as the United Kingdom, where they represent 10 per cent. They are very likely to remain at this relatively marginal level.

The sector is no longer production-led but marketing-led. Traders must be able to capture information downstream and pass it back to the production end via new orders for growers to be able to adapt to changing demand. This raises the problem of power control within agricultural organisations of the first stage of marketing.

Marketing Services from Producer Countries are Supporting their Products with Promotional Budgets for their Trademarks

Origin	Trade marks
Agrexo	Carmel
Israel Citrus Marketing Board	Jaffa
New Zealand Apples and Pears	Enza
Holland	Greenery
South Africa	Outspan and Cape
Belgium	Flandria

The wholesale market sector has suffers much damage from integrated distribution, especially large-scale distribution, which has moved away from the wholesale markets. Large catering firms and catering the wholesale markets. Large catering firms and catering suppliers, however are tending to deal directly with producers or importers. Wholesale firms offer a better service than wholesale markets, hence surplus capacity, fragmentation and keen competition.

Concentration of Distribution

European Union: According to a French study the average market share of integrated distribution is 60% in 1992 and 35% for Belgium. Onifhlor (France) indicates that the increase of the concentration is approximately 3% per year. In Spain the number of hypermarkets (over 20.000 m^2) doubled between 1987 and 1992. The special attention of the large grocery distributors for those products is explained not only by the need to present a complete fresh offer to the consumer but also because fruit and vegetables are traffic builders.

This evolution is not without consequences for the internal sector. It leads to a concentration of demand which is often criticised by producers and their cooperatives. In general the large distribution chains:

- have strong negotiation power
- buy big volumes of uniform quality normalised and standardised
- have their own marketing policy on the basis of in house brands
- are cost aware
- favour concentration in international co-operation

Fruit and vegetables account for about 10 per cent on average of food turnover of hypermarkets andut 10 per cent on average of food turnover of hypermarkets and supermarkets in the EU. Purchasing by hypermarkets and supermarkets is certain to continue rising in future.

The share of integrated distribution is currently 45 per cent and 35 per cent approximately of available quantities of fruit and vegetables, respectively. It may exceed 50 per cent by the year 2000. The situation varies considerably from country to country. Germany is the leading country, with integrated distribution accounting for over two-thirds of available quantities, compared with only one-quarter in Spain, one-tenth in Italy and a marginal proportion in Greece.

The importance of large-scale distribution for these products can be explained not only by the commercial objective of covering all consumer purchases but also because a good fruit and vegetable department attracts more customers to the store and encourages them to come more often and more regularly and therefore boosts the turnover for all departments throughout the store.

Competition between big names is steadily increasing. There is also more pressure on prices since the problem is how to offer goods at cheaper prices than the competitors, hence the use of trademark products and special offers. Supermarket chains and names are continuing to merge: the two leaders account for about one-quarter of the integrated trade turnover in Germany and three-quarters in Switzerland. European leader and three-quarters in Switzerland. European leaders are emerging, with the tendency to develop stations grouping European supply upstream of sales points. The limitations of such a strategy are already apparent, however, in particular for perishable goods and with regard to the trade-off between low prices and guarantied physical, chemical and organoleptic quality.

This trend has an impact on the entire sector. Demand is becoming more concentrated although this is often highly criticised by producers and their associations. As a rule, buying agents for supermarkets:

- exert considerable power to lower prices in negotiations;
- try to obtain large quantities of product of uniform, standardised and regularised quality;
- operate on a just-in-time basis thus passing a large share of stock management over to producer groups and local dealers.

Ranking of the Ten Leading European Distributors by Turnover (1.000.000) ECU

Name of the company :	Turnover 1995	
Metro	32.900	
Railway	25.900	
Carrefour	24.800	
Intermarche	29.400Intermarche	29.400
Edeka	22.200	

United States: In 1994, 74 per cent of retail food store sales were controlled by integrated chains: this share was 62 per cent in 1974 and 58 per cent in 1954. The balance was accounted for by independents, both affiliated and unaffiliated. At the turn of the century virtually 100 per cent of retail food sales were made by unaffiliated independents while the current share is only 1 per cent.

The dominance of the chain stores is even greater if all grocery stores are included, as opposed to just supermarkets. Integrated distribution now represents 81 percent of total food sales.

Distribution is increasingly controlled by larger operators, which have acquired smaller firms, and the turnover per firm is steadily rising. There are approximately 400 integrated wholesale-retail headquarters buying offices in 1996.

The largest fruit and vegetable consumption channels in the United States are:

- retail food stores (59 per cent);
- hotels, restaurants and catering establishments (40 per cent);
- direct farmer-to-consumer sales (1 per cent).

Larger bulk purchases are efficiently handled by direct sales and distribution from the shipping point rather than through terminal markets. Many chains put their ping point rather than through terminal markets. Many chains put their own personnel in the production region to ensure product quality and availability.

Integrated wholesale-retail buyers use wholesale markets only for small batches of exotic or highly perishable produce. Thus integrated distribution by-passes the wholesale market stage and avoids intermediate handling and services costs. Direct dispatching of produce to the integrated distribution centre avoids any break in the cold chain and ensures improved quality.

The trend toward concentration is clear is the United States and Europe and is steadily rising all over the world. There are three main consequences.

Concentration at the supplier level :

- competition between chains as fruit and vegetables are major traffic builders and consumer awareness for health and quality is likely to increase;
- further requirements in terms of logistics, treatment, packaging and product range.

Large-scale distribution is continuing to concentrate markedly and overall, its market share is expanding. The specialist fruit and vegetable trade is becoming increasingly concerned with quality and is adapting its logistics, services and image.

Opportunities and Constraints

Quality Standards

Quality standards form a system of classification which helps to make the market more transparent by indicating certain characteristcs the market more transparent by indicating certain characteristics of a product marketed. Standards reduce transaction costs and allow trade to develop without physical attention to the product.

What would happen if standards were no longer compulsory?

Initially there would be little change. Big traders would continue to refer to the standards as they really need them for their transactions, but the system would gradually break down on local and regional markets, reducing market transparency.

Standardisation has also been used as an instrument of market management. For example, it led to two measures which have been the subject of some debate in EU :

- the ban of the marketing of class III (low quality) products on the fresh market, and
- the setting of the minimum size for Community apples of the large-fruit varieties in 1990 at 65 mm (used in international standards). The primary reason for this was the desire to limit supply on the market by eliminating small apples.

Using standards for market management has certain advantages:

- since it applies to products for market and for withdrawal, it limits the quantities available on the market without pushing up subsidised withdrawals. In situations of surplus (as is the case for many products) it contributes towards market balance, reduces budget expenditure and improves producer prices;
- in some cases (*e.g.* kiwi fruit) it is theucer prices;
- in some cases (*e.g.* kiwi fruit) it is the only available market management instrument.

But these advantages are sometimes ill perceived by consumers, who are deprived of produce suitable for consumption and traditionally consumed in their region.

As for end distribution, objective competition is very likely to be brought into play for brand ownership or "quality signs" between production and large-scale distribution. Although this seems obvious for stabilised products, for fresh produce sector management is a prerequisite for any reliable quality approach. In any case, large-scale distribution buyers:

- will be steadily fewer in number;
- will request supplies 12 months ahead with an increasingly broader product range in order to meet consumer expectations;
- will tend to take on the supply function upstream although this is not part of their job, especially for perishable produce.

Alongside large-scale distribution, the specialist trade stands its own ground, especially if it can become structured to guaranteed product quality and thus asserts its special role in the segment of the markets identified, based on consumer recognition of their competence.

Product Differentiation

According to this strategy market and market segments are selected on the bases of:

- size of the market

- expected growth of the market
  - o size of the market
  - o expected growth of the market
  - o consumer preference
  - o structure of retail
  - o relative strength with respect to competing suppliers

To build a dominant position in the market a brand policy is needed with strong brands offered to the consumer. Elements contributing to a strong position of trade are:

- o quality
- o the power to offer large quantities
- o variety of offer
- o service
- o reliability
- o meeting logistic services demands of the consumer.

In this way fruit and vegetables can become an attractive partner in the management of the customer, which is the bases of a long lasting relation of cooperation. It is clear that the more developed the economy is, the bigger the demand for differentiation in stead of uniform products.

Integrated Production

There is no doubt that we have the duty to consider the health of the consumers in the framework of the strategic options. Maximum residue limits for pesticides have to be commonly agreed on and respected. Therefore there is an urgent need for:

- o international harmonisation of pesticides
- o international harmonisation of residue limits
- o harmonisation of control methods of residues.

This harmonisation should apply both for the production as well as for imported products.

Integrated production is a tool allowing production of fruit and Integrated production is a tool allowing production of fruit and vegetables in which the residues of chemical products are extremely low and where the production respects as much as possible the biological equilibrium of the environment. This means not only the satisfaction of the general demand of consumers who like to consume natural products. Two factors, the hygienic and sanitary quality on the one hand and the equilibrium of the environment on the other hand are closely related and interdependent.

By the study of positive and negative interactions of animal and plant-based elements it is possible to reduce the need for additional chemical products and in some cases the chemical products can be replaced by biotechnological means which are absolutely safe for the environment.

The interaction of both factors are the key for integrated production.

Home Market

It could be a paradox but because of market enlargement the competition becomes more and more global and at the same time the home bases become smaller and not less important. Porter indicates that Europe becomes more a collection of home markets who with the help of their own national and regional authorities fight a heavy internal competition. He stresses that a cluster, an interdependent system of supply and support activities has to be created in which the structures of commercialisation have a unique importance. In this sense Pmercialisation have a unique importance. In this sense Porter illustrates it by indicating the worldwide success of cut-flowers from Holland. Porter indicates that, as compared to the more sunny southern countries, Holland has not the same comparative advantages. But it is precisely the building up of a network in this sector with many supplying and supporting activities that Holland became the centre and is no doubt the world leader. Because the international success greatly depends on the existence of such a network of home bases, the unilateral product thinking has to be changed to a product market thinking. This emphasizes the need to produce a high quality product rather than a cheap product, in order to succeed.

The commercialisation has an irreplaceable role to fulfil. There is a need for a synchronized relation back and forward between the system of production and high quality marketing. The importance of such a network of production and trade with local production as a starting point for success on international markets can not be stressed enough. Therefore one should not rely on government where not enough initiative and experience are available.

Integration

Promotion is one of the four key elements of the Kotler strategic approach. Promotion is certainly not the starting point for a product strategy. But in a competitive environment the power of promotion in a competitive environment the power of promotional campaigns can not be underestimated. As for the building of a dominant position in the market a brand policy generally is needed. It supposes that brands will be established that appeal to consumers and are supported by intensive publicity campaigns.

The wholesale trade is an alternative to highly concentrated distribution centres for fresh produce in the context of trade growth.

The future of the wholesale trade depends on its capacity to adapt traditional functions to new services and new customer requirements. If it is able achieve this, it can offer an alternative to the supply centres of large-scale distribution buying offices. Wholesale markets need to be sufficiently large and have a suitable range and calendar of supplies. They must also acquire logistics. However, they will still need to show brand leaders that their interest can be met, which seems to be the case for perishable goods.

Acquisition of shareholders in groups or networking are carried out by:

- major international firms with activities mainly based on trade in storable produce (*e.g.* bananas and citrus fruit) and produce from the southern hemisphere (apples, kiwis, pears and grapes) :
- national firms trying to grow to an appropriate size;
- wholesale firms diversifying supplies and integrating more and more activities.

Performance

Downstream firms (large-scale distribution, wholesalers offering full services) constantly have to find ways of meeting consumer requirements via their sales points. The constraints include the following:

- planning deliveries to stores according to quantity, quality, uniform products, product ranges and services (timeliness; observance of temperature needs etc.);
- setting competitive prices with minimum variations (the problem facing distributors is not to purchase at the lowest price but not to buy goods at higher price than the next distributor, for a product of equal quality);
- product differentiation through own brands (distributor brands).

In order to meet these constraints downstream firms must:

- match production potential with market requirements;
- organise relations between operators through partnerships or any other form of solidarity between operators in a sector strategy, especially for implementing effective qualitative strategies;
- be able to conduct experiments, demonstrations, information, transfer of know-how and training;
- offer commercial and advertising and promotional economic services.

At the production and dispatching stage, geographical concentration in the most favourable areas with the best fruit and vegetable growers is creating specific production regions.

These are uni creating specific production regions.

These are units with specific soil and climate characteristics but also economic entities and spaces where the various operators share common interests, whether they be producers, dispatchers or processors.

This is the level on which applied research and experimentation should focus and where the cropping practices and capacity to develop new products and new varieties make sense as the prerequisites for any systems.

In North America, integration is liking producers in several areas, so that stores and consumers are be assured of year-round supply of commodities. In this regard, producers and marketers operate in several states or even countries, getting information on production and handling techniques wherever it can be obtained. In general, it is much more difficult for universities and experiment stations to show as much flexibility in providing information to growers and marketers as these client groups need.

Diversity of Horticultural Crops

The North East region has its own unique combination of living species, habitats and ecosystems, which together make up its diversity rich resource. While speaking strictly about plant diversity, two regions of the country are termed as hot spots. These are – Western Ghats and the North Eastern hill regions. In all living organisms the species is the single most useful unit to use in diversity assessment. Species richness and the relative abundance of different species is another criterion to measure the degree of diversity. The number of endemic species also reflects into account while assessing the richness of diversity.

The North Eastern region of India comprising eight states namely Arunachal Pradesh, Assam, Manipur, Meghalaya, Mizoram, Nagaland, Tripura and Sikkim has vast physiographical variations, which have been represented in 6 agro climatic zones. North East region is one of the richest reservoir of genetic variability and diversity of different crops i.e. various kinds of fruits, different vegetables, spices, ornamental plants and also medicinal & aromatic plants. The diversity for horticultural crops of this region has mainly been managed by local farmers, often women.

Considerable diversity exists among the regional horticultural species including variation in plant type, morphological and physiological characteristics, reactions to diseases and pests, adaptability and distribution. Apart from the nutritional value, many regional horticultural crops are used for medicinal purposes, income generating and poverty alleviation programmes in the rural areas.

Exploration and Distribution

Fruits

Citrus: Being the home of several citrus species, rich genetic diversity occurs in the region. In lemon alone, as many as 32 strains are available.

The species *C. limon, C. medica, C. jambhiri, C. ichangensis, C. latipes, C. macroptera, C. assamensis, C. indica* and *C. aurantium* are considered indigenous to this region. The Indian wild orange *C. indiaca* is found in the Naga hills and Meghalaya.

Banana and Mango: Maximum genetic variability of *Musa acuminata* and *M. balbisiana* occurs in NE India. *M. flaviflora* is localized to Manipur and Meghalaya. There are other species found in Sikkim and Khasi Hills, which need systematic collection and conservation. Some native *Mangifera spp.* are found in Tripura, Manipur, Mizoram and South Assam. Wild form of *M. indica* and its allied species *M. sylvetica* occur in Arunachal Pradesh, *M. khasiana* and *M. pentandra* in Assam.

Temperate Fruits: Rich diversity occurs in Pyrus, Rubus, Ribes and Prunus. The Shillong plateau of Khasi hills in Meghalaya has many Prunus species such as *P. napalensis*, *P. undulata* and *P. cerasoides*. *Pyrus pyrifolia* var. cubha makai (*P. serotina* Red) is grown semi commercially in Meghalaya, Manipur and other places. Wild kiwi (*Actinidia callosa* and *A. stragosa*) is found growing in the natural forests of Arunachal Pradesh and Sikkim.

Tropical and Sub-tropical Fruits: A large number of other tropical and subtropical fruits belonging to the genera Garcinia, Artocarpus, Phyllanthus, Annona, Averrhoa, Persia, Aegle, Passiflora, etc., are *ENVIS Centre, GBPIHED* 2 found growing wild in the region. One of the indigenous fruits that requires attention is jackfruit, which grows abundantly in Tripura, Assam and Meghalaya with a large number of cultivars.

Under Utilized Fruits: Of the 300 edible plant species found in the NE region, some of them are really worth consuming by various ethnic groups of tribals. Two species of *Elaegnus*, viz *E. latifolia* and *E. pyriformis* are known to be grown in NE region (Pandey, 2002). It is quite common in Sibsagar (Dikho valley),

Naga hills, Khasi and Jaintia hills. *Docynia indica* and *D. hookeriana* are commonly found in the region. *Pyrus pashia* is a medium sized deciduous fruit tree also found in NE region.

Canavalia Gladiata

Leafy Vegetables: The important leafy vegetables include lai (*Brassica juncea*), lafa (*Malva verticillata*), palak (*Spinacea oleracea*). In addition to these a wide variety of indigenous leafy vegetables are also available. These are amaranth (*Amaranthus spp*), puroi sag (*Vasella rubra* and *B.alba*), sorrel (*Rumex rasicarius*), etc. Other indigenous leafy vegetables used occasionally are jilmilsag (*Chenopodium album*) and Kalmou sag (*Ipomea reptans*). *Amaranthus viridis*, *A. lividus*, *A retroflexus* and *A. spinosus* are important leafy types grown in North East India (Sarma, 2001).

Tuber and Rhizomatus Crops: Based on the colour and the skin to broad types of sweet potato are grown in the region. These are the white skinned and the red skinned variety. A number of dioscorea species *alata, bulbifera, brevipetiolata, esculenta, hamiltonii, hispida, kamaonensis, nummularia, pentaphylla, puber* and *quinata* were recorded in the Region. *D. hamiltonii* occurs in humid forests of

NE hills (Sarma, 2001). Varieties of tapioca or cassava like M-4 and hybrids H-97, H-165, H-226 have also been cultivated to some extent. In colocasia also there is a wide variability even with in one species such as *Colocasia esculenta* (Sarma, 2001).

Lesser-known Vegetables: In addition to the above there are a large number of indigenous vegetable crops that are used particularly by the tribal population. Maximum such varieties are available in Arunachal Pradesh. Tree bean (*Parkia roxburghii* G. Don.) is one of the most common of multipurpose tree species in Manipur and Mizoram (Kumar *et al.*, 2002). In the hilly areas tree tomato (*Cyphomandra betacca*), a perennial shrub producing red tomato like vegetables, is also grown and used as such. It is grown as backyard venture crop in Meghalaya (Thakur *et al.*, 1988). Another vegetable tree, growing in the lower altitude zones and popular among the people, is drum stick or horse reddish locally called Sajina (*Moringa oleifera*).

Cho-Cho (*Sechium edule*), a native of tropical America, is a very popular vegetable in the region commonly called squash and grows abundantly without much care and attention. *Flemingiavestita* known as Sohphlong, is consumed raw. It is a weak climbing/trailing type, under ground tubers, distributed the humid to sub tropical regions of NE India upto 1500m (Sarma, 2001). Kakrol (*Momordica cochinchinesis*) and kartoli (*M. dioica*) are widely spread in Assam, the Garo hills of Meghalaya (Ram *et al.,* 2002).

Spices: A wide range of variability has been found in ginger and turmeric. Lakadong the local variety of turmeric is grown in Jowai area of Meghalaya. Poona and Thinglaidon performed better in most of the states except Tripura and Sikkim indicating suitability for large scale cultivation for green ginger production. Nadia variety of ginger is well adopted in North East region. A type of ginger having rhizomes with bluish black tinge inside called Black ginger is grown by the inhabitants of Mizoram. Wild relatives of large cardamom (*Amomum subulatum*) and cinnamon are available in the forests of this region.

Ornamental Plants

The ornamental plants, which are found in NEH region, include species of Mannolia, Rhododendron, Cassia, Erythrina, Calustemom, Dacasenda, Myrica, Bauhinia, etc. About 600 species of orchids are naturalized in great proportion in this region. Wide range of variability has been found in Sikkim, Arunachal Pradesh, Meghalaya and Manipur.

Medicinal and Aromatic Plants

Medicinal plants of known use like *Rauvolfia serpentina, Solanum khasianum, Dioscorea proceri, Coptis teeta* are widely available in the region. Orchids as medicinal agent have been used by the tribals of NE states like Arunachal Pradesh, **Nagaland**, Meghalaya and Mizoram.

Germplasm Evaluation

In order to facilitate effective utilization of plant genetic resources, it is important that the plant genetic resources are evaluated for productivity including components, crop duration, resistance to biotic and abiotic stress and quality of produce. A dwarf and late maturing polyembryonic mango cultivar Moresh bears sweet fruit with high pulp content within 2 years from planting and is free from stone weevil. The locally available specieslike *Malus baccata* is widely used as rootstock of apple while *Pyrus pashia* is a common rootstock of pear. The ripe fruits of *Elaegnus*, viz *E. latifolia* and *E. pyriformis* are edible and used for making refreshing drink (Pandey, 2002). The fruits of *Docynia indica* and *D. hookeriana* acidic, greenish with red tinged are eaten fresh and in pickles as well as in jelly preparation.

Some of the brinjal varieties have excellent quality in having large size, soft flesh and less seeds. An important species of medicinal importance (solasodine content) is *S. khasianum*. Another species *S. torvum* is extensively used in Ayurvedic medicine system. Three tomato varieties namely Manileima, Manikhamnu and Manithoibi were released by State Variety Release Committee, Manipur and found suitable for rice-based cropping system. *L. pimpinellifolium* is also good source of resistance to late blight and tomato leaf curl virus (Seshadri and Srivastava, 2002). In chilli, a collections from Tezpur (Assam) has been found to have the highest capsaicin content recorded so far anywhere in the world.

Cho-Cho (*Sechium edule*) produce large starchy edible roots in addition to fruits. The National Bureau of Plant Genetic Resources, Regional Station, Shillong is maintaining more than 10 germplasms of Indian bean. Winged bean has excellent nutritional qualities particularly being very rich in protein (Rao and Dora, 2002). In the trials conducted by ICAR Research Complex for NEH Region, Umiam varieties of sweet potato such as S-162, S-69 and S-30 have been found to be promising. Recent studies indicate the dioscorea hybrid H-312 and H-1687 are more adapted to wide range of conditions ensuring higher tuber yield. Tree tomato is consumed as delicious chutney when raw or after roasting and peeling off the skin. The tubers of *Vigna vexillata* are rich in carbohydrates and minerals. The tubers of *Flemingia vestita* are rich in iron (2.64 mg), phosphorous (64.06mg) and contain fair amount of protein (2.99 g), calcium (19.77) and carbohydrates (27.02g) (Sarma, 2001).

Turmeric variety Megha turmeric-1 (earlier known as RCT-1) and ginger variety Nadia were found suitable for the region. Lakadong is a turmeric variety found in this region, which has high percentage of curcumin content (7.4%).

The native species of orchids having ornamental value and market potential usually belong to Aerides, Anachnantha, Arundina, Cymbidium, Dendrobium, Paphiopedillium, Phaius, Renanthera, Phycostyllus and Vanda etc. (Borthakur, 1992). The juice from cut leaves of *Cymbidium giganteum* is used by Khasis for clotting of blood while the juice of Vanda flower is used as eye drop for cure of glaycoma. There is immense scope of medicinal and aromatic plants for possible use of local germplasm in the improvement of vegetable crops for the use in horticultural industry (Borthakur, 1992).

Conservation of Diversity

Land Tenure Issues: Land tenure systems vary widely among different North-Eastern states, which are quite different from the rest of India. The complexity in land ownership and tenurial rights makes it difficult for survey, demarcation and consolidation of land. Therefore, cadastral survey and land demarcation are completely absent in the hill areas of northeast.

Gender and Equity Issues in Natural Resources and Diversity Management: Unequal distribution of land resources is responsible for increasing dependence on forests by certain sections of the society leading to diversity degradation. Resolving the gender and equity issues concerning natural resource management is equally important in North-East as in the other parts of the country.

Inter-departmental Coordination: Inter-departmental coordination is essential to the sustainable management of horticultural resources in the region.

Smuggling of Timber Across the International Border: The illicit felling of trees and timber smuggling across the international borders has been the most important cause of horticulture areas/forest degradation in border.

Shifting Cultivation: Unregulated shifting cultivation by the local tribal populations has been a major threat to sustainable diversity management particularly in unclassed and community forests of the region.

Inter-state Border Dispute: There exist a lot of inter-state border disputes among the north-eastern states. Most of these border areas are forest lands and because of boundary disputes, such lands are often declared as 'no man's land' and hence, do not come under any form of management. This leads to the degradation of diversity in such areas.

Insurgency: The long insurgency problem in some states such as Assam and Tripura has considerable impact on diversity conservation.

Prospects

Although there are not many agencies/organizations working exclusively for diversity conservation in northeast *per se*, the activities taken up by many organizations including non-governmental and traditional institutions, government departments and scientific institutions have direct or indirect implications for diversity conservation.

State Government Agencies: Many state agencies are now involved in such diversity conservation activities as establishment of germplasm banks for horticultural crops.

Research Organizations: Many state and central government research organizations including universities of the region are engaged in research, inventory and conservation of diversity in theregion. Such organizations are Botanical Survey of India, Shillong, G.B. Pant Institute of Himalayan Environment and Development, (Northeast Unit, Itanagar and Sikkim Unit, Panthang), Indian Council of Agricultural Research for North-Eastern Hill Region, Barapani, Shillong with campuses through out the northeast, State Forest Research Institute, Itanagar, NBPGR, Shillong, North-Eastern Hill University, Shillong, Nagaland University, Kohima, Mizoram University, Aizawl, Arunachal University, Itanagar, Tripura University, Agartala, Assam University, Silchar, Tezpur University, Tezpur, Guwahati University, Guwahati, Assam Agricultural University, Jorhat, Regional Research Laboratory, Jorhat, Dibrugarh University, Dibrugarh.

Non-Governmental Organizations: Many non-governmental organizations are now working for the conservation of diversity in northeast, although most of them are local and grassroots level.

International Donor Agencies: International donor agencies in Meghalaya, Manipur and Assam, and Nagaland have been playing crucial role in conserving the diversity through their respective projects.

International and National Policies and Conventions: All the international treaties and national policies have significant impact on the conservation of diversity in the northeast.

Academic Institutions Including Schools and Colleges: The educational curriculum in the universities, colleges and schools has an important role to play in diversity conservation.

Shifting Cultivators: The shifting cultivators and other traditional farming communities of northeast have played a key role in conserving the rich horticultural crops germplasm of the region. In spite of the availability of many hybrid and high yielding varieties these farmers have been cultivating the traditional varieties for generations.

Gaps

The depletion of diversity and inadequacy in actions to conserve the diversity of the region may be attributed to several factors, which range from inadequate knowledge about diversity and its components to adoption of wrong and inappropriate policies by the concerned stake-holders.

Gaps in Knowledge and Information: Information on urban diversity is scanty Species inventory in inaccessible areas of Arunachal Pradesh, **Nagaland**, Karbi Anglong and North Cachar hills of Assam, and parts of Mizoram and Manipur is yet to be made Information on genetic diversity is extremely poor Information on wild ornamentals and aromatic plants is scanty.

Gaps in Vision: Most of the programmes and activities being undertaken by the state governments are shortsighted. Long-term planning based on sustainable development strategies and integration of diversity conservation issues with development planning is the need of the hour.

Monoculture Plantations: In order to increase the revenue generation, the State **Horticulture** Departments pursue the policy of raising plantations of commercially important species by clearing and burning the natural diversity areas.

Introduction of High Yielding Varieties/hybrids of Crops: The **horticulture** departments are introducing various high yielding varieties/hybrids of horticultural crops. This is associated with increasing use of inorganic fertilizers and chemicals for plant protection. Such policies not only ignore the indigenous species and varieties but also have adverse effects on existing flora and fauna.

Conversion of Natural Vegetation Areas into Cash Crops: Cash crops such as tea, coffee, cardamom, ginger and turmeric are being encouraged by several agencies throughout the northeast. Many shifting cultivation rehabilitation programmes introduce these cash crops as alternatives to shifting cultivation. These plantations are obviously taken up on lands with natural vegetation (in past or present) at the cost of diversity.

Gaps in Policies and Legal Structure: The wrong conservation policies with focus on economically important species have been harmful to diversity. Such policies as adopted in Tripura, Mizoram, **Nagaland** and Meghalaya have not only decreased the species diversity in natural/rehabilitated forests but have also resulted in accelerated soil erosion and loss of soil moisture. The policy of rehabilitation of *jhumias* through rubber plantation as has been done in Tripura may prove to be a disaster for other floral species in such areas.

The policy of promoting high yielding varieties and assessment of progress and success on the basis of consumption of fertilizer and plant protection chemicals has led to ignoring the indigenous varieties. The government subsidy and credit policy is instrumental in adopting these schemes.

Through the Public Distribution System only HYV are distributed. There is a need to include distribution of indigenous varieties too. The planners have not considered the role and value of diversity in preparing developmental plans. Such ignorance has been responsible for taking no efforts to conserve and enhance diversity.

Most of the problems are related to increase in population. The rate of population growth in the northeast is unusually high. This causes tremendous strain on the natural resources and adoption of certain policies that are not very friendly to conserve diversity. No population policy has been adopted for future planning.

Education policy does not include teaching on diversity conservation. The school curriculum should be able to mould the young minds in favour of diversity conservation. There is no policy as such to create awareness among masses for diversity.

Gaps in Institutional and Human Capacity: The number of trained taxonomists in the region is grossly inadequate. This is one of the most important bottlenecks for completing the inventorization of diversity.

Not all persons concerned with management of genetic resources understand the concept of diversity in proper perspective. Many of them suffer from biased attitudes. So it is imperative that those who plan, decide and implement the developmental programmes are adequately trained and educated in favour of diversity conservation.

There are a number of institutions, departments, colleges, universities, NGOs, local community groups that follow certain programmes having bearing on genetic resources. While framing their programmes, these agencies are motivated to pursue their own goals in watertight compartments without considering their impact on other programmes or existing resources. There is no institution, which can make them sit together and discuss the programmes in a holistic manner.

Gaps in Diversity Related Research and Development: Regeneration and cultural practices for many species need to be researched and standardized for their cultivation. Threatened species need immediate action for ensuring their continued existence. Identification and classification of threatened species need to be done.

Richness of diversity of horticultural crop species is yet to be fully inventorized and documented. There is a conspicuous gap between research and field needs. The established formal institutions like university departments, departmental research stations and others rarely consult the farmers and local communities about their problems while pursuing research. Need-based research needs to be encouraged.

Conclusion

Considerable diversity exists among the regional horticultural species including variation in plant type, morphological and physiological characteristics, reactions to diseases and pests, adaptability and distribution in the NE region. Apart from the nutritional value, many regional horticultural crops are used for medicinal purposes, income generating and poverty alleviation programmes in the rural areas.

Problems relating to diversity conservation and development of horticulture in north eastern region are land tenure issues, gender and equity issues, inter-departmental coordination, smuggling of timber across the international border, shifting cultivation, inter-state border, insurgency, etc., which are responsible for horticulture diversity degradation. The establishment of NBPGR, ICAR, BSI, and various universities in north east has made tremendous impacts in collection, evaluation, conservation and utilization of regional germplasm for development of horticultural varieties in this region. Keeping in view the regional demand for horticultural crops more germplasm needs to be identified for collection particularly for high yield, quality, resistance to diseases and pests, tolerance to frost and acidity.

Infrastructure

The provisions of adequate infrastructure is a per-requisite for industrial development. The State will therefore endeavour to provide necessary infrastructure to facilitate the growth of industries. The concept of Build-Operate-Transfer shall also be considered in identified areas of Power Generation, Road & Bridges, Warehousing and Industrial Infrastructure either in the private sector or Joint Venture with the State Government. Existing infrastructure shall be strengthened/upgraded to suit the ever increasing demands of the industrial sector.

1. *Communication:* The length of National Highway in the State is 471.17 Kms and the State Highway is over 1050 Kms. All the major towns and villages are now connected by all weather roads. The State is connected by Rail and Air through Dimapur. Efforts will be made to convert the roads leading to all industrial zones, industrial estates, Growth Centre, EPIP, etc. to sustain heavy vehicular traffic for transportation of raw materials and finished products.
2. *Power:* The present peak demand of power in the State is only 50MW. With commissioning of more 132/66KV substations in Wokha and Tizit during the 2000, augmentation and System improvement the peak demand is expected to reach 80 MW. Even after commissioning of the 75 MW Doyang HEP, 24 MW Likhimro HEP and 24 MW Thermal Power Station, the State will still be in shortage of power. Therefore there is tremendous scope of investment opportunities in the Power sector. Nagaland has potential of more than 2000MW in hydel power generation.

Industrial Infrastructure: The following industrial infrastructures are available for setting up industrial units in the State.

a. Old Industrial Estate and New Industrial Estate at Dimapur.
b. Infrastructure projects at the final stage of completion.
   i. Mini Industrial Estate at Tizit (Mon) and Kiphire (Tuensang).
   ii. Growth Centres under 'No-Industry District' at Longleng and Noklak (Tuensang).
   iii. Growth Centres under DIC schemes at Chuchuyimlang (Mokokchung), Baghty (Wokah) and saring (Mokokchung).

a. Projects under implementation and expected to be ready for use by 2001-02.

i. Industrial Growth Centre, Ganeshnagar, Dimapur (Under PM's new initiative).

ii. Export Promotion Industrial Park (EPIP), Ganeshnagar, Dimapur.

a. Industrial Zones:

The following areas in the State has been identified as Industrial Zones where industrial units set up will be eligible for the package of incentives provided under the industrial policy: Besides the above, the Government will endeavour to notify identify similar zones in other potential areas such as Naginimara-Tiru & Jalukie valley for a balanced industrial development of the State. The Government will also set up IID Centres Food Parks, IT Parks and chain of Cold storages etc. in the State to provide all necessary industrial infrastructures to facilitate prospective investors/entrepreneurs to set up industries in the State.

Tele Communication: All the district headquarters are now well connected and the Tele Communication Deptt, is in the process of expanding its network to cover all the recognized villages in the State. The State Government is also planning to set up Community Information Centre in Block Headquarters and also open more new Telephone exchanges in backward area for training data base and application oriented end use. Internet facilities are available in Dimapur and Kohima, Laying of optic fibre cable to enable Video conferencing facility is available at Kohima and is expected to be commissioned soon at Dimapur. (4) *Finance:* Poor C:D ratio in the North East in general and particularly in Nagaland has been one of the major hurdles in the industrialization of the State. The State has a reasonable banking network in place. All the major banks have branches in Dimapur and Kohima, while other districts are served by the SBI and other State level banks. In order to provide adequate credit linkage to the industrial sector, the Nagaland Industrial Development Corporation Ltd. (NIDC) which is also acting as SFC shall be strengthened to cater to the needs of the entrepreneurs in the services of IDBI, SIDBI, NEDFi, NMDFC,

NSDFC and commercial banks for channeling credit for industrial development of the State.

Resource Base of the State

a. *Agro Forest:* Nature has endowed the State richly in the form of rainfall and fertile soli and a salubrious climate, through which the agro/forest resources are abundant. Nagaland has been the major exporter of round logs, veneers, plywood and sawn timber to the rest of the country till the ban imposed by the Supreme Court. 'Tree farming' has been taken up on a massive scale by the last 10 to 15 years and it is expected to be a major resource for industrial activity in the near future.

b. *Medicinal & Aromatic Plants:* Due to the favourable agro climatic conditions, Medicinal Herbs and Plants are abundantly available in the hilly regions of the State. Citronella/Eucalyptus oil production in the State is presently taken up by the Government in the form of demonstration units. Scientific harnessing of this potential in the private sector can yield good results. One of the most valuable medicinal plants commonly known as Ginseng is also available in Nagaland which has tremendous commercial potential for export.

c. *Horticulture:* As per the recent study taken by the Government of India, the North Eastern Region has been identified as a potential region for promotion of Horticulture because of the favourable agro-climatic conditions. At present, 42000 hectares of land is covered under horticulture crops but this area can be easily quadrupled given proper pre and post harvest infrastructure. Food Processing Industries based on organic horticulture produce, therefore, have tremendous potential. The State has a comparative advantage in the availability of fruits such as pineapple, plum, bananas, passion fruit and citrus fruits, bamboo shoot and other exotic spices. Introduction of appropriate technology and an efficient marketing network coupled with necessary quality control measures can lead to industrial boom.

d. *Animal Husbandry and Meat processing:* The State has high potential for diary and allied products as well as for processing of hides and skins for leather Industry. At present, scientific processing of Poultry/Piggery and other meat products is lacking in the State. Therefore, there is potential for setting up modern abattoir and other meat processing units in the State. Komul brand of diary products have received a good market response in and outside the State. There is huge scope for production and marketing of other value added diary products. Rearing of Angora rabbits for "Angora wool" is also a potential which can be exploited.

e. *Sericulture:* The agro-climatic condition is very favourable for the development of Sericulture in the State. The State Sericulture department is implementing various Sericulture schemes. There are 34 Sericulture farms which provide technical and other supports like training of farmers, distribution of improved varieties of saplings/ seedlings/supply of disease free layings (DFLs) etc: The Govt. has identified Eri Culture as the trust area for development of sericulture in the State by bringing 200 acres of land under Eri culture in the next three years. Through this the production of cocoons is targeted to yield 80 MT.

f. *Floriculture:* Floriculture can be taken up as a very viable industry in the State. The wide variety of orchids available in the State can be exported to earn valuable foreign exchange. Many species of orchids are available in Nagaland, which has tremendous commercial potential. There is scope to grow exotic and new varieties of orchids through tissue. With the declaration of Guwahati as international Airport, cut flowers can also be exported from the State.

g. *Mineral Resource:* Nagaland is rich in minerals resources. The following mineral reserves have been established to date(i) petroleum and natural gas, (ii) nickel cobalt-chromium bearing Magnetite, (iii) marble, dimensional/ decorative stones, (iv) coal etc. The huge reserves of more than 1000 million tonnes of high chemical grade limestone in the eastern Nagaland is a major prospect for setting up of industries such as;

i. Cement
ii. Calcium carbide
iii. Bleaching powder
iv. Hydrated lime

Other industries which could be set up in this sector are;

i. White & green marble mining and polishing.
ii. Ceramic glazed tiles.
iii. Ceramic crockery.
iv. Ceramic insulators.
v. Slate for building materials.

This high grade limestone can be exported for foundry, blast-furnace and other metallurgical and chemical industries.

Tourism: The picturesque landscape of the hilly state of Nagaland and its manifold cultures and customs offer vast potential for development of Tourism Industry in the State. Each of the 16 major tribe have their own distinctive colourful costumes, festivals, dances and music. Nagaland is aptly called *'land of festivals'*. The rich art and culture of the people is a major tourist attraction. The State is blessed with salubrious climate throughout the year and one can visit it anytime. The Government of India's decision to release the RAP will open new opportunities of development of tourism industry. Such opportunities exist in eco-tourism, adventure tourism, anthropological tourism, and ethnic cultural tourism.

Human Resource: Although the literacy rate of the State is 83% which is much higher than the national average, the State still lacks in skilled manpower. This scenario is slowly changing as large number of Naga youths are going out of the State for higher and technical education. In addition to the ITI centres run by the government, private training institutes, particularly in computer education are also coming up. This requires a major push so as to develop and up-grade the skills of the local people by utilizing the service of reputed training institutions/organizations. Development of managerial capability of local youth shall be taken up through intensive EDPs. The State shall encourage private investors/organizations to set up training Institutions in the State. Since English is the official language of the State, this added advantage can be used effectively by any training institute.

Thrust Areas

The following shall be the thrust areas for Industrial Development :

a. Food Processing Industries
b. Tourism Industry
c. Agro-based industries
d. Mineral based industries
e. Handloom and Handicrafts
f. Sericulture
g. Floriculture
h. Electronics and IT
i. Pharmaceuticals
j. Petrochemicals
k. Bio-tech Industries

Strengthening Direct Industries Centres: The District Industries Centres (DIC) play a vital role in the development of SSI and Tiny sector Industries. The Government shall strengthen the DICs with adequate facilities and also train the officers &staff to keep pace with the latest technology/skill. The newly created Dimapur DIC will be upgraded to full-fledged DIC at the earliest. Each DIC will act as principal agency to guide the entrepreneurs in matters of setting up of Industries including selection of industries/entrepreneurs, preparation of project and arrangement of finance for projects.

NRI & Foreign Direct Investment: The State Government will actively promote foreign direct investment in the projects that involve and improve infrastructures like power, roads, social and health care facilities. Similarly, foreign direct investment shall be encouraged in manufacturing activities in thrust areas identified under this policy.

Trade & Export: The Government will take effective steps to provide necessary in infrastructure in all the identified trade centres at Longwa, Pangsha, Mimi, Molhe and Avangkhu on priority. The Government of India for expediting formalization of trade with Myanmar in order to bring up the status of the trade centres on par with Moreh-Tamu sector in Manipur. The Government of India has declared Guwahati airport as an International Airport with all clearance facilities but it is to be made functional. NIDC, the only declared Export House in the State, shall be further strengthened. The Government shall also take up with the Government of India to upgrade the cargo handling facilities at Dimapur airport. The potential for export of Naga handloom and handicraft products, processed fruits and flowers (fresh & dry) will be fully exploited.

State Public Sector Undertakings: In the backdrop of the ongoing process of liberalization and privatization, the public sector enterprises, especially in the non-core sectors, are losing their relevance. Therefore, in keeping with the present trend, the government will make a distinction between the promotional and purely commercial PSUs. While the promotional undertakings will be strengthened to fulfil their role effectively, the viable commercial PSUs will be restructured so as to run them on professional lines even by bringing in professional management through lateral induction. The general policy would be to progressively reduce financial and budgetary support to all the PSUs so as to make them stand on their own feet. The Government will not hesitate to close down the unviable sick PSUs.

Congenial Atmosphere for Industrial Development: Peace is a pre-requisite for creating congenial atmosphere for any developmental process. Therefore, the Government of Nagaland shall strive to create a congenial atmosphere for sustained growth by maintenance of law and order. The State Government shall consider provisioning a dedicated state force for industrial security.

Package of Incentives

In tune with the industrial policy resolutions, the government of Nagaland has formulated a package of incentives for promotion and setting up of industrial units and revitalization of sick industrial units in the State. The details of the incentive packages are under:

Period of Validity: The new package of incentives, hereinafter referred to as "Incentive Scheme-200", shall become effective from the date of issue of notification of the policy and remain in operation for a period of five years or till such time as the Government may consider fit and proper. The Government also reserves the right to make amendments to the schemes.

Effective Date: Effective date for the Incentive Scheme-2000 shall be the date of issue of notification of the policy.

Definitions

Ancillary Industry means an industrial unit which is engaged or is proposed to be engaged in the manufacture or production of parts, components, sub-assemblies, tools or intermediates or rendering of services, undertaking supplies or renders or proposes to supply or renders not less than 50% of its production or services as the case may be, to one or more other industrial undertaking and whose investment in fixed assets in plant and machinery whether held on ownership terms or on lease or hire purchase, does not exceed Rs. 100 lakh.

Eligible Unit means only new units set up on or after the date of issue of notification of the policy and existing units undergoing expansion/modernisation/diversification the same place in the State of Nagaland provided that :

The location of the Registered Office shall be within the State of Nagaland, (relaxation may permitted with approval of the Industrial Facilitation Committee *in special cases)*

Existing Unit means an industrial unit which is or was in commercial production at any time prior the date of issue of notification of the policy shall be considered as an existing unit for the purpose of the Incentives Scheme-2000.

Expansion of an industrial unit means additional fixed capital investment exceeding at least 25% of the capital investment of the existing unit. For the purpose of calculation, gross value of all the capital investment made on land, building and plant and machinery of the existing unit will be taken into consideration. Expansion shall also imply an increase of at least 25% in the existing installed capacity as well as increase of additional employment of at least by 10%. Prior to going for expansion, the unit should be operating at least at a minimum of 80% capacity during the three previous years and prior intimation to the concerned implementing agency for Industrial Policy-2000.

Diversification means separately identifiable investment made by an existing industrial unit in the fixed capital assets to set up a project for manufacturing of new product(s) provided that the additional investment in the fixed assets is not less than 25% of the gross fixed capital and increase of additional employment is at least by 10%.

Modernization means separately identifiable investment made by an industrial unit involving new/improved technology having a definite advantages in reduction of cost of production provided the additional investment in fixed assets should not be less than 25% of the gross fixed capital. The incentives available to the unit undergoing expansion/diversification/ modernisation shall be for the additional investment made and/ or the increase in production over the average of three previous years prior to the year when the unit goes for expansion/ diversification as applicable. The year for this purpose shall mean a financial year. The unit which has taken up expansion/ diversification/modernisation plan prior to coming into force of this policy should intimate the concerned agency within three months from the date of implementation of the policy.

100% Export Oriented Unit means an industrial unit which undertakes to export its entire production of goods subject to relaxation as permitted by Government of India from time to time.

Finished Products means and includes, the item manufactured by the eligible unit as considered under the project/scheme approved by the concerned term lending agency and or by the concerned term lending agency and or by the implementing agency, together with by-products/scrap, which may get generated as incidental to and during the main production activity. Fixed Capital means and includes cost of Land, Building including Factory shed, Godown, Laboratory, Plant & Machinery, Installation Charges and Pre-operative expenses capitalised, Electrical and such other equipment which are directly related to production activities.

Government means the Government of Nagaland.

New Unit means all an industrial unit which has commenced commercial production on or after the date of issue notification of the policy.

Non-Eligible Unit means all industries which are mentioned in Schedule-1 to this policy as non-eligible.

Raw Materials shall mean and include components, intermediate goods, substances, consumable stores required in the process of manufacture and packing materials utilised for packing of finished products by an eligible unit.

Relief Undertaking Unit means an industrial unit declared as a relief undertaking by the Government of Nagaland or units declared sick by State Government/director of Industries & Commerce under any other scheme/Act of the Government.

Schedule means a schedule to this scheme.

Small Scale Industry (SSI) means an industrial unit where the investment in plant and machinery does not exceed Rs. 1.00 crore.

Small Scale Service Business Enterprise means an enterprise where the investment in plant and machinery does not exceed Rs. 25 lakh.

Tiny Industry means and industrial unit where the fixed capital investment dies not exceed Rs. 25 lakh.

Export Oriented Unit (EOU) means a unit that exports at least 35% of the total production.

State means the State of Nagaland.

Eligibility:

i. All industries except those at Schedule-1 shali be entitled to assistance under the in Incentive Scheme-2000.

ii. Incentives under Incentive Scheme-2000 shall be available to eligible units in the private sector, joint sector, State public sector and the co-operative sector. Central public sector undertakings shall not be eligible for any incentives under the scheme.

iii. All new units shall be eligible for the Incentive Scheme-2000.

iv. Relief undertaking/sick units shall be eligible for the incentives under Inventive Scheme-2000 for a maximum period of 3 (three) years.

v. The unit undergoing expansion/diversification/ modernisation shall be eligible for incentives only for the additional investment made *i.e.* for the expanded/ diversified/modernised portion of the existing unit.

Eligibility Certificate:

i. Eligibility Certificate is a certificate which shall be issued by the *Industrial Facilitation Committee* for Medium & Large Industries, Directorate of Industries & Commerce for SSI Unit and District Industries Centre for SSSBE/ Cottage & Tiny sector. This will be issued after ensuring that all the criteria for eligibility have been fulfilled to the full satisfaction of the concerned *industrial Facilitation Committee.*

ii. No right or claim for any incentives under the scheme shall be deemed to have been conferred by the scheme merely by virtue of the fact that the unit has fulfilled on its part the conditions of the scheme and the incentives/ subsidies cannot be claimed as a matter of right.

iii. The incentives under the scheme cannot be claimed unless the 'Eligibility Certificate' has been issued under the scheme by the Implementing. Agency concerned and the unit has complied with the stipulations/conditions of Eligibility certificate.

iv. The decision of the Implementing Agency, subject tot such directions as Government may issue from time to time in this regard shall be final and binding.

Certification of Employment: The Certification of Employment would be given jointly by the Deputy Commissioner and the District Employment Officer of the District where the industrial unit is located. Where the Office and production centre of industrial unit are located in more than one district, the Director of Employment and Craftsmen Training will issue the certificate of employment.

Implementing Agency: The Implementing Agency for the Incentive Scheme-2000 in respect of the Large & Medium and SSI sector shall be the Director of Industries & Commerce and for Cottage & Tiny and SSSBE shall be District Industries Centres.

Procedure for Disbursement of Incentives: Separate guidelines for issuance of eligibility certificate along with the application forms and procedures for disbursement of incentives would be issued.

Priority for Disbursement: The disbursement of the Incentives by the Implementing Agency shall be in accordance with the chronological order of approved claims. Priority shall be however given to SSI units and 100% Export Oriented Units in thrust areas. Preference will be given to units where at least 50% of the employees are local tribal youth.

Interpretation: The decision of Industries & Commerce Department, Government of Nagaland, as regards interpretation in this policy resolutions/incentive scheme shall be final. The State Government reserves the right to declare any more incentives or new guidelines or amend any provision(s) including withdrawal of any of the incentives/subsidies as and when necessary for development of industries in the State from time to time under the provision of the Policy.

Right of the Government: The State Government reserves the right to review the matter regarding sanction/disbursement of subsidies/incentives to the eligible industrial unit(s) and in this connection the State Government's decision shall be final.

An Introduction to India's Northeast: If the history of world civilization can be described in terms of the gradual diminution of the significance of geography, the 1990s should be regarded as the most important decade in the history of mankind. Two developments make it so – the integration of the world economy spearheaded by the WTO and the emergence of Information Technology. India was one of the first among the developing countries to recognize the opportunities and challenges thrown up by the new forces of world integration. In 1991 the Government of India set into motion a wide–ranging reform agenda, which has acquired national acceptance already. India, happily, has also found itself more than ready for the IT revolution because of its vast trained manpower resources.

But even as we are speaking of the integration of the world economy, there is at least one region of India that is yet to be fully integrated even with the national economy. I am referring to the northeast of India, joined with the rest of the country by a narrow corridor, aptly called the chicken's neck. Land-locked and geographically isolated, this region has continued to stagnate, though myriad studies have confirmed its great economic potential. In this presentation I shall briefly outline the broad features of the region's economy, chiefly for the benefit of prospective investors.

The Region: The Northeastern Region (NER) comprises the states of Arunanchal Pradesh, Assam, Manipur, Meghalaya, Mizoram, Nagaland and Tripura, with about 8 per cent of the country's geographical area and about 4 per cent of its population. The region is known for its physiographic, ethnic and linguistic diversity. It is one of the most bio-diverse zones of the world. Some 75 population groups inhabit the area, speaking something like 200 languages and dialects. For convenience the people are usually divided into three groups – hill tribes, plains tribes and others. The tribes constitute only about one fourth of the population though they are distributed over large areas and four of the states are almost entirely tribal. The plains are inhabited by Hindus and Muslims, by and large. Christians are the majority in the hill states of Meghalaya, Mizoram and Nagaland. In spite of the heterogeneity, the region is quite homogeneous in other respects. The tribal societies have an egalitarian milieu. The Hindus do not observe the caste system at all rigidly. Restrictions on inter-dining are unknown. Except in some parts of Assam and Tripura, the type of economic disparity and poverty that is found in the rest of India is also not in evidence here.

The population of the region was estimated at 34.79 million in 1995 of which Assam alone accounts for 24.61 million. Arunachal Pradesh has the lowest population of less than a million. The average population density is 123. But this can be misleading because of the wide diversity. Assam and Tripura have a fairly high population density of 286 and 263 respectively. Arunachal has the lowest density, just 10. Others range between 33 and 82. The decennial growth of population in all the states has been higher than the national average. It has been argued that immigration from Bangladesh and Nepal is an important contributory factor for the high rate of population growth.

Literacy is higher than the national average in all states but Arunachal, with Mizoram boasting of near 100 per cent literacy. The valleys in Assam and Manipur are economically more active in the basically agricultural economy. The hills and foothills, though comprising two thirds of the total geographical area, are not equally active. Except in the plains, land is controlled by the community through a traditional system of land tenure.

Unbalanced Economy: There are differences among the seven states of the region with respect to their resource endowments, levels of industrialization and infrastructure facilities. Yet on the whole all these economies are underdeveloped, agrarian economies with very weak industrial sectors and inflated service sectors. The industrial sector has mainly developed around tea, oil, timber and mining. Only the tea plantation industry employs a large labour force. In Assam it employs more than 500,000 workers. State-sponsored industrialization in the form of sugar mills, jute mills, paper mills and even food processing units has not been successful. Therefore, the economy of the region is still primarily agricultural. But the full potential of even this sector is yet to be exploited. This is because of primitive farm practices of slash and burn, shifting cultivation (Jhum) in the hills and mainly single-crop, traditional farming in the plains. As a result, the region is not able to produce even adequate food grain to feed its population. The states of the region import food items worth Rs.20 to Rs.25 billion annually from other parts of the country.

Since agriculture and industry have not really taken off, the pressure for employment is on the service sector. As a result, this sector has expanded disproportionately. Due to low economic activity, the states of the region are resource deficit. But a large portion of these limited resources is spent mainly to maintain the service sector. Except in Meghalaya and Nagaland, the contribution of agriculture to Net State Domestic Product (NSDP) is much higher in the region. Mining is an important contributor to the Meghalaya economy. Similarly, forestry and logging play an important role in Arunachal and Nagaland. But after the Supreme Court's restrictions on tree felling, the future of this sector is in doubt. Compared to all India average of 20 per cent, manufacturing's contribution is between 5 and 8 percent in the region. The contribution of construction is high in the region. In Arunachal it is 20 percent and in Nagaland, it is as high as 31 percent of the NSDP. Public administration and other services play a very important role in the region. Compared to the national average of 11 percent, its contribution in the region ranges from 15 percent in Assam to more than 40 percent in Tripura.

Employment is mainly in agriculture and the service sector. In Assam, Meghalaya and Nagaland about three fourths of the work force is engaged in agriculture.

Economic Growth: In the eighties, the economic growth rates were higher than the national average in Arunachal and Nagaland. And, except in Assam, they were quite comparable to the national economic growth in other states. This was quite natural as the region was starting from a low level of economic development. It would also be interesting to know the performance of these economies since the 1991 reforms. There was severe contraction at the national level in 1991-92. But, by and large, there was no visible impact on the Northeast economies. They did not experience the stabilization impact. Later, when the national economy started recovering fast, these economies started slowing down. They have clearly failed to take advantage of the national expansion.

Agriculture: Agriculture, with allied activities, is the mainstay of the NER economy. Despite the large geographical area, the land available for cultivation is limited. The area of agriculture operations is confined to only 5 per cent of total geographical area in Arunachal Pradesh and 8 per cent in Manipur. It is less than one third of the total area in Mizoram and Tripura. Compared to All India average of 64 per cent, area under agricultural use is between 40 and 50 per cent in Assam, Meghalaya and Nagaland. Further, out of this limited area, only a small portion is irrigated. It is about 15 per cent in Assam and Arunachal, and 12 per cent in Tripura. In Nagaland and Meghalaya the gross irrigated area is slightly higher. Only in Manipur, the gross irrigated area is about 40 per cent of the cropped area, which is higher than the national average of 36 per cent. But the unique feature of the region is its high rainfall, which ranges between 3,000 and 12,000 mm. In the hilly areas, jhum cultivation continues to be a major agricultural practice adding to soil erosion and destruction of forests. The average size of operational holdings is smaller than the national average in Assam, Manipur and Tripura. But in other states, due to jhum cultivation and common ownership, it is larger.

Except Arunachal Pradesh, no other state in the region is self sufficient in food grains. Barring Manipur, the consumption of fertilizers and improved seeds is very low. Consumption of electricity in agriculture is picking up only in Tripura. In 1995-96, the food grain production in the region was more than 5 million tonnes. Food grain yields were higher than the national average in Manipur, Mizoram and Tripura. Food grain productivity has increased substantially in Manipur and Mizoram in the last couple of years. But it has declined in Arunachal. Rice is the principal crop of the region. Rice yield per hectare is highest in Manipur followed by Tripura. Both are above the national average. Compared to that, yield of rice is only 72 per cent of the national average in Assam. Apart from Assam, Tripura also produces tea. Other States such as Meghalaya and Arunachal are also experimenting with tea cultivation successfully.

Since the agro-climatic conditions of the region are congenial for horticulture and spice cultivation, most of these States produce a wide range of such crops as fruits, nuts, vegetables, spices, plantation crops and aromatic and medicinal plants. The major horticulture crops grown are banana, oranges, pineapple, coconut, jackfruit etc. In 1995-96, NE region produced about half of the total ginger production of India. Major contributing States were Meghalaya, Arunachal and Mizoram. Other spices such as chillies, turmeric, cardamom etc. are also grown. Potatoes, sugarcane and oilseeds are important too. Rubber plantation is becoming important, particularly in Tripura. In animal husbandry, the region produces only about 20 per cent of its meat requirement and about 15 per cent of the egg requirement. In 1993-94, the NER produced about two hundred thousand tonnes of fish. But this was way behind the region's requirements.

Industry: As in March 1994, the region had 166 large and medium scale industries. Out of these industries, more than 70 per cent were in Assam. Some of them have already been declared sick. There is hardly any large or medium private industry outside Assam. And even within Assam, about half of the industrial units are located in only three districts-Kamrup, Tinsukia and Dibrugarh. On paper, there are about 40,000 small scale units in the region. According to government statistics, half of them are sick. This indicates that disproportionate emphasis on small industry may also not be the right approach in the region. In fact, there will not be more than a few thousand units actually operating.

Gross output per employee as well as fixed capital per employee is less in the region compared to all India averages. This indicates either low fixed capital or over-manning. Net value added is high in States having oil and gas or mining industries like Assam, Meghalaya and Tripura. But it is negative in Manipur.

Percent distribution of value of output by manufacturing industries in the NER indicates that most of the industrial units are in food processing, wood and wood products, non-metallic mineral products, rubber, plastic, petroleum and paper. In Assam and Tripura about half of the units are in the food sector. In Nagaland 71 percent are in wood products. In Meghalaya, 64 per cent are based on mineral products. Outside these four or five areas, there is hardly any industry. Overall, the region contributes only a little over 1 per cent to the value of the output of the manufacturing sector in India.

Infra-structure: The infrastructure index for the states of the northeast has been worked out by the Tenth Finance Commission appointed by the Government of India as follows (against an all India index of 100): The low scores are partly explained by the fact that the hill states have little or no rail links. Except Meghalaya, which is surplus in power, all other states are short of their requirement of power. The region has about 6 percent of national roads and 13 percent of national highways. Road communication is thus not much behind the national average. The Guwahati airport in Assam has been developed as an international airport. All the state capitals of NER will be connected by air services in the next two or three years. Most of the NER states have also made good progress in telecommunications. Most of these states are much above the national average in terms of schools and teachers.

Human Resource: One point on which the region scores over most other states of India is its high literacy rate. Female literacy in the region is significantly higher than the national average. In most of the states English is the medium of instruction. Educated young people are usually fairly fluent in written and spoken English. This is perceived now as an advantage that the region enjoys over many other parts of India in the age of IT-enabled services. The NER has 35 institutions of higher and technical education. The capital of Meghalaya, Shillong, has always enjoyed the status of a major centre of education in India and attracts students not only from other states of the northeast but also from places outside the NER. Unemployment among the educated is lower than the national average but still quite high.

Potential, Prospects and Investment Opportunities

Geographical Advantage: Contrary to traditional wisdom, the NER is increasingly being viewed as having a location advantage, having extensive borders with Bhutan, Bangladesh, Myanmar and China. In addition, these states are very close to the dynamic East and Southeast Asian economies. There are some studies, which reveal that movement of people and merchandise between NER and China, Burma and Thailand did exist in the past. In recent years trade routes to Myanmar have been reopened and traditional trade with Bangladesh has been increasing. As of date, the estimated value of trade with Bangladesh is Rs. one billion and that with Myanmar is of the order of Rs. two billion. In addition a considerable volume of cross border trade, formal and informal, with Myanmar and Bangladesh is in existence. The value of informal cross border trade has never been properly assessed; but it is believed to be worth much more than the official trade.

The Government of India has already made a small beginning for expansion of trade with neighbouring countries. Trade talks with Bangladesh and Myanmar are on. The Government of Meghalaya has set up an export promotion industrial park at Bynihat, about 20 km from Guwahati, with central assistance. The Central Government is also setting up a trade fair exhibition centre in Guwahati. There is a proposal to set up an NER trade promotion council. Opening up of old trade routes with China is also under consideration. The Guwahati airport has been upgraded as an international airport, as already stated. There is a possibility that a trans Asian railway linking Myanmar and Bangladesh will go through the NER. All this is good augury.

Natural Resources: The NER has oil reserves of over 500 million tonnes, natural gas reserves of 190 billion cubic meters, coal deposits estimated at over 900 million tonnes, limestone reserves of about 5000 million tonnes, clay deposits of more than 11 million tonnes and granite deposits of 2 million cubic meters. It has 49 percent of its total geographic area under forest cover. The hydel power potential of the region is 32000 mw at the most conservative estimate, one third of India's total potential. Two thirds of the region is covered by hills of great natural beauty, an ideal resource for adventure and eco tourism. The forests of northeast abound in various kinds of medicinal plants and there is a rich folklore about the healing properties of various plants.

Food Processing: There is a tremendous scope for food processing industries in almost all the states of northeast.

These industries will be based on local resources as well as local demand. As for local resources, the region grows a wide variety of horticultural crops and spices. I have mentioned them earlier in this presentation. These products find markets far and wide in Indian cities and in Bangladesh. Unfortunately, there are few fruit processing units worth the name. This has acted as a dampener on farm production and productivity.

Tea is becoming a very popular crop in the entire region. Whereas Assam alone was the tea growing state in the past, at present almost every other state has experimented with this plantation crop with success and more and more tea gardens are coming up in so-called non-traditional areas. This crop could afford very attractive investment opportunities along with tea processing.

Demand-based food processing industry has great potential in animal products. Nearly the entire population of the region is meat eating. On the other hand, rearing of animals for meat is neglected and there are no meat processing industries at all.

Tourism: The whole of NER is one of the most beautiful areas in the world. It is home to one of the richest varieties of flora and fauna on the globe. The bio reserve in Meghalaya is regarded as the richest one known. The Kaziranga National Park in Assam was once adjudged the best small wild life sanctuary in the world. There are numerous such sanctuaries spread all over the northeast. Opportunities for trekking, rock climbing, hang gliding, angling etc. are practically inexhaustible. There are several golf courses, the most famous one being the one in Shillong. Cultural, pilgrimage and conference tourism have very attractive possibilities too. Tourism in this region requires a great deal of investment, which, unfortunately, has not been forthcoming so far.

IT Industry and IT Enabled Services: There are many reasons why IT industry and IT enabled services should be an attractive investment opportunity in the NER. The excellent climatic conditions of the hills, pollution free environment, high literacy rates, a large body of trained manpower, proficiency in English amongst the educated, and huge investments proposed by the Government of India into IT related infrastructure together constitute a favourable environment for investment in this sector. Proximity with the countries of south East Asia is also an added advantage for this purpose.

Government of India Initiatives: Since 1996 successive Prime Ministers of India have taken a personal interest in formulating a special package for the NER to promote economic development, the most recent one being a Rs. 10 billion package announced by Prime Minister Atal Bihari Vajpayee. These initiatives cover the following general ground:

- Identifying gaps in infrastructure and basic minimum services with central investment to remove them.
- Generation of employment.
- Earmarking of 10 percent of each central ministry's budget for specific programmes in NER.
- Separate industrial policy for NER.
- Separate export promotion policy for NER.
- Improved credit flow to the region.
- Improvement and expansion of railway services.
- Water management and flood control measures.
- Border trade and development of export trade routes.
- Massive investment in telecommunications and electronics.
- Tourism

In 1997 the central government declared a separate industrial policy for the NER. Among other things, this policy declared specified industrial areas as excise and tax-free zones for all kinds of industries. For industries based on local raw material the same benefits have been made available in any part of each state of the NER. The new policy has brought about a marked change in the industry's perception of NER as an investment destination. From 1998 onwards Indian industry has shown a great deal of interest in this region. More than a hundred proposals have been cleared by the respective state governments, with many more in the pipeline. In the designated tax-free zones new industries are coming up at a very encouraging pace.

The State of Meghalaya

I shall like to conclude this presentation with a few words about my own state, Meghalaya. Meghalaya has everything going for it, economically speaking. It is one of the few states in India that is surplus in power. Financially, it is one of the best-managed states in the country and has been recognized as such by the Government of India. It has a massive 4200 million tonne limestone deposit, coupled with a fairly impressive 550 million tonne coal deposit. It has other minerals such as granite, sillimanite, china clay, fire clay, phosphorite and quartz. It has a salubrious climate and a dust and pollution free atmosphere, itself a resource for high tech and service industries. It grows oranges, pineapple, peaches, pears, plums and bananas in abundance. It also grows plenty of ginger, turmeric and other spices. Finally, it is a beautiful tourist destination with opportunities for golf, nature and eco tourism, adventure tourism and caving.

The Government of Meghalaya has identified agro-based and food processing industries, mineral-based industries, electronics and IT industry, hydel and thermal power generation, orchid cultivation, spices, oleoresin and other essential oils, medicinal plants, tea and healthcare & recuperation as the thrust areas for investment.

One reason I can think of why Meghalaya has not attracted much private capital in the past is that the state was created to safeguard tribal interests. For years after its creation the government and the people were preoccupied with this perceived need to protect the people from outside pressures. In recent years there has been a major shift in attitudes. Both the government and the people have come to the realization that the state can grow and develop only with massive private investment. In 1997 the state government declared its Industrial Policy with several incentives and an aggressive agenda for attracting outside capital. This initiative has been supported by the Government of India with a special package of incentives if its own. The combined effect is that Meghalaya has, for the first time since its creation in 1972, started receiving attention from the industry. Today the multinational company Lafarge has indicated its interest in setting up a cement plant in Meghalaya. There is a Rs. 600 million potato processing plant under construction. An Indian cement company, Saurashtra Cements, has already entered the state. A number of other industrial projects are in the pipeline.

The Government of Meghalaya has decided to set up a new township near Shillong to meet the growing needs of urbanization and industrialization. It will be a Rs. 8 billion project. We are very keen on foreign and PIO involvement in consultancy services and financing for this project.

Before I close, I shall like to make a suggestion to the GOPIO. The organization should consider opening up a regular communication channel with NER by setting up an office at Shillong. The region is looking to people of Indian origin for advice, guidance and finance to realize its true potential and to find its place in the sun.

Tripura Agriculture Department

Tripura, being endowed with fertile soils, abundant moisture and sub-tropical climate offer immense scope for production of a wide variety of tropical and sub-tropical fruits and vegetables. However, the horticultural activities practiced in the State are mostly backyard farm activities, characterized by less efficiency, and low level of productivity, though it holds, promise for quantum leap both in terms of production growth and market expansion. It was, therefore, a felt need to exploit the horticulture development potential of the State in a scientific and systematic way, in a perspective time-frame to generate sustained levels of income and employment opportunities.

It was also contemplated in the document, "Approach to peoples plan in Tripura" that the production of fruits and vegetables should be doubled and to this end, all available land and water resources will be put to appropriate use. In pursuance to the above directive, a draft perspective plan for development of Horticulture in Tripura for the period 2002-2012 was prepared by the Directorate of Horticulture & Soil Conservation and was approved by the council of Ministers, Tripura on July 3, 2001.

The Strategies adopted for Preparation of this Plan are:

Identification of specific Horticultural crops with reference to sustained higher economic return.

Selection of areas for cultivation of such crops based on potential for production, access to the market, level of farmer's skill and experience etc. preferably in cluster basis.

Provision of improved technology and production input.

Provision for post harvest management.

Provision for required storage and marketing support.

Provision for crop-specific training support to the farmers & extension functionaries.

Exploration of new fields viz. Floriculture, Mushroom cultivation, Tissue culture, Organic farming etc.

The Salient features of this plan are as follows:

- Identified Horticultural crops like orange, pineapple, mango, banana, litchi, jackfruit, papaya, cashewnut, arecanut, coconut, black pepper, ginger, chilli (capsicum), betelvine, summer cabbage, summer cauliflower, seed tuber from TPS and horticultural crops like flowers, orchids, enthurium, vanilla and mushroom have been included.
- An area of 12,750 hectares (net) and 18,965 ha (gross) will be brought under the above crops during ten year plan period.
- Substantial increase in productivity level of identified crops will be attempted through adoption of improved technology and supply of quality planting materials.
- Increase in the production of fruits from the present level of 3.0 lakhs ton to 4.05 lakh tons and, vegetable from present level of 3.25 lakh tons to 4.38 lakh tons by the terminal year of the plan period.
- For transfer of technology to the farmers, crops specific on campus training of farmers and extension staff at the rate of 1500 Nos. and 250 Nos. per annum, respectively will be undertaken at HRC, Nagicherra, besides taking

a group of 80 farmers per annum to area as of latest Horti developments outside State.

- Use of Bio-fertilizer & vermin compost will be popularizes and adoption of Integrated Pest Management (IPM) techniques will be encouraged.
- The additional requirement of cold storage is expected to develop under private sector/cooperative sector for which necessary assistance is proposed to be made available under various schemes of NABARD & NCDC.
- The Food Park is proposed to be created by the Department of Commerce & Industries, Government of Tripura. Small scale processing of horti crops is also proposed to be encouraged through mobile processing unit as per programmes of Govt. of India. Besides the exising capacity of TSIC is proposed to be increased from the present level of 100 tons to 400 tons of finished products.
- For improving domestic as well as export market of fresh fruits, intervention of NERAMAC has already been started.
- Role of financial Institutions in extending necessary assistances in relevant cases has been well schemed.

Tripura is a small hilly and landlocked State in the North East Region of India with very poor communication facilities. The economy of the State is basically agarian and more than 70% population depends on Agriculture for their livelihood. The State of Tripura with a geographical area of 10,491 sq. km. is predominantly hilly. It is surrounded on all sides by deltaic basin of Bangladesh except for a small part in North East which adjoins Karimganj District of Assam and Mizoram.

Location: Tripura is situated between 22°57¢ and 24°82¢ North latitudes and 91°10¢ and 92°22¢ East longitudes with the tropic of cancer passing through it. Tripura is by and large a hilly terrain with small hills and hillocks cress-crossing the valleys.

Climate & Rainfall: The State of Tripura enjoys a typical monsoonal climate with variations ranging from sub-tropical to temperate conditions in hilly areas. The rapid change in topography results in significant climate changes within a short distance. The climate of Tripura exhibits a strong seasonal rhythm. The year being divisible to four characteristic seasons namely Winter (Dec-Feb), Premonsoon (March-April), Monsoon (May-September) and Post monsoon (Oct-Nov). The Monsoon period lasting for about 5 months from May to September is the longest season of the State. June is the wettest month followed by July & May. The average annual rainfall in the state is 2100 mm. The maximum and minimum temperatures during winter are 27°C & 24°C respectively. The maximum and minimum relative humidity for the atmosphere (at 10.00 am) are 85% & 57% during the months of July and January respectively.

Soils: In general soils of Tripura are classified into two categories – upland & lowland General fertility of the soil of the State is medium. The soil varies in reaction from very strong to strong acidic with medium organic matter content and low available phosphorus and potash contents. The $_{p}H$ varies from 4.85 to 5.8 Soil texture varies from sandy clay loam to clay loam. Ground water potential is good in valley fills, dissected mounds and certain part of denudational hills. The organic carbon content is medium.

Land Use Pattern: The total area under forest being 6,06,168 ha (about 58% of the total geographical area), forestry makes the dominant land use class in the State. It is followed by the Agriculture. The areas under different land uses are:-

Forest	6,06,168 ha
Under Non-Agri Use	1,33,500 ha
Under Misc. tree crops	27,151 ha
Net Cropped Area	2,350 ha
Net Cropped Area	2,80,000 ha
Total Geographical Area	10,49,160 ha

Traditional Crafts

Handicrafts: Excellent variety of handicrafts using bamboo and cane are made by different ethnic groups which have earned great name and fame throughout the country. The most famous handicraft products are Room Divider, Decorated wall panels, Attractive furniture of cane, Different decorative pieces using Bamboo roots, Bamboo Dining table mats, Floor mats and various other gift items. Tourist can watch the craft persons at work in different villages and buy handicraft and handloom products from Purbasha, a Govt. of Tripura undertaking Sales Emporium and other private Sales Emporium throughout the state.

Cane and Bamboo Handicrafts: From time immemorial Tripura has carved out a name for itself in the field of Handicrafts. The Gifted artisans produce wonderful objects of crafts from simple material like cane, bamboo & wood. There rare artistic skill has not been streamlined in the manufacture of exquisite household pieces. Tripura's unique topography and the gracious nature bestowed their choicest blessings on the hereditary artisans of Tripura. Here three distinct cultures *viz.* Hinduism, Buddhism and Islam have converged together to give shape and content to a unique tradition that found eloquent expression through immemorable work of art & crafts made out of very simple materials like cane, bamboo, clay, wood, palm leaf etc. With the passage of time, there has been changes in the demographic character of the state. But in each phase of her history, Tripura has shown remarkable inner strength of assimilation of synthesis, while retaining her own traditional heritage. With the original distinct tribal motifs were added the skills of Manipuri and Bengali artisans who came subsequently to settle in this land. Time has changed and so also the quick adaptability of our gifted artisans who never failed to respond to the demands of contemporary tastes.

Cane & Bamboo occupy a distinctive place in the life of Tripura. From cradle to grave, there is hardly any occasion, complete without the use of cane & bamboo.

Handloom: The handloom industry of Tripura is very famous. At present it is not organised. The weavers are scattered all over the state. As the unemployment rate is very high in the state, and there is no dearth of literate youth, the state government is developing the Handloom Industry into a full fledged industry. Besides giving a better value to the weaver for their work, it will also generate employment for educated youth. The development of this Industry will directly result in the upliftment of the rural mass development of Tripura.

The Handloom of Tripura in two categories:

Non-commercial looms are operated by the tribal inhabitants of the state. They weave fabrics for their own consumption whereas the commercial looms are controlled by the Bengalis as well as Manipuri weavers. So the Handloom Industry of Tripura is the reflection of Tribals, Manipuri and Bengalies.

However, handloom is the single largest and perhaps the oldest industry in the state of Tripura and is scattered in the rural as well as semi-urban areas. It has grown with the civilisation itself and people in this region look to weaving inseparable part of life. The industry has grown as the art has passed from one generation to another. It plays a dominant role in the economic development of the rural masses. This industry also forms an important sector providing employment to the rural poor.

The tribal women of Tripura religiously devoted to weaving. The beginning of ritual has always been with the offering of flowers and beautiful piece of Riha and Risa given by the great grand mother and followed by each and every tribal family as their tradition.

In fact it was a pre-requisite for every tribal girl to know weaving. Otherwise she should not be considered as a bride. The tribal people produce their own clothes imparting them various designs, texture as well as colour; all of which are long lasting. Among the traditional produces, mention may be made of Risa and Riha (breast garments). Riha is remarkable for its colourful variety and texture.

The Bengalee weavers who have migrated in Tripura from the erstwhile East Pakistan (Bengal) are the main commercial weaver and play a significant role in the development of Handloom Industry. They have also contributed to the development of texture, variety in the Handloom Industry. The Manipuri weavers are also playing an important role in developing Handloom Industry in Tripura.

Horticulture: Tripura is land of high hills, hillocks, interspersed with rivers and valleys having moderately warm and humid climate, well distributed annual rainfall of 2500 mm. The terrain soil and climate of the State of Tripura are ideally suited for rain-fed Horticulture. Fruits like Pineapple, Jackfruit, Orange, Litchi, Cashewnut, Coconut, Lime, Lemon are produced in abundance. Fruit crops in the State are grown on hillocks, availing the well distributed rainfall without application of any fertilizers or chemicals.

Tripura is fortunate to have agrciultural as well as climatic conditions suitable for production of a large variety of horticultural crops chief among which include "Queen" and "Kew"; varieties of pineapple, oranges, lichis and cashew. Each of these from Tripura are well known in the rest of the country. These products are also free of any chemical residue. Insipte of this, the Fruit Processing industry in Tripura is at a nascent stage. The Government has helped the state of Tripura in pineapple processing-NERAMAC (a Government of India undertaking), has set up a modern Pineapple Juice Concentrate Plant at Nalkata in North Tripura District, with a processing capacity of about 5760 TPA. The Plant at Nalkata is operating at a low level due to non-operation of aseptic filter. However, the same is being rectified and natural juice concentrate will be available from the pineapple season of this millenium year.

Rubber: Recently declared the "Second Rubber Capital of India" by the Rubber Board, the state of Tripura has availability of good quantity of high quality rubber offers ample scope for setting up of rubber-based industries in the state. Rubber has therefore been identified as one of the thrust areas in Tripura. Studies show that about 100,000 hectares of area in the state can be brought under rubber plantation though the area under rubber cultivation at present is estimated to be about 23,500 hectares, second largest, after Kerala. The yield @ 1,500 kg per hectare and the quality of rubber are also comparable to Kerala's plantations. In fact,

The state government has taken up an ambitious programme to increase the area under rubber plantations by another 20,000 MT by the end of Ninth Plan *i.e.*, by 2001-02 AD, with assistance from the Central Government, the Rubber Board and the World Bank. As a result, it is expected that the rubber production, which is presently about 5,000 MT per annum, will increase to about 20,000 MT.

Natural Gas: Tripura has vast reserves of natural gas in non-associate form. The gas is of high quality, with high methane content of about 97% and without any presence of H_2S of Sulphur and other impurities. ONGC has been actively engaged in exploration activities in the State since 1972. The availability of superior quality natural gas, at concessional price, offers a great opportunity to prospective investors, to set up gas-based industrial units, using natural gas as Feedstock.

The Global Citrus Germplasm Network

The Global Network on Genetic Resource Conservation and Utilisation of Citrus and Citrus Relatives was formally constituted, under the aegis of the FAO, to be addressed to all countries interested in citrus germplasm conservation and utilisation. The network is functioning on a voluntary basis and it is involving national institutions as well as existing regional and inter-regional networks dealing with citrus genetic resources conservation and utilisation.

In this respect, the existing regional and inter-regional citrus networks (The Inter-American Citrus Network-IACNET; The Network on Citrus Improvement for the Wider Mediterranean Region-MECINET) and those under constitution (Asia-Pacific and Sub-Saharan Africa) and also the CLAM, NESCRAF and any other regional networking initiatives that are dealing with citrus germplasm, are actively participating in the Global Citrus Germplasm Network.

The scientific activities can be summarised as follow:

- o develop better general morphological descriptors evolving from existing systems through wide consultation (current system being limiting for non-orange citrus);
- o develop standard non-morphological descriptors appropriate for citrus (includingological descriptors appropriate for citrus (including guidelines to bring in information on resistance to biotic and abiotic stresses)
- o Supply descriptor system to global computerised information system working group.

- o Define uses of markers (molecular and biochemical) according to needs and applications
- o Develop information coding system of marker methods suitable for data bank use
- o Determine system requirements that provide adequate information with global coverage and worldwide access to the information.
- o Develop system to on-line stage.
- o Develop worldwide coordination and operation mechanism for the information system.
- o Identify taxa which are insecure or under-represented in collections as priorities for *in-situ* characterisation and conservation efforts.
- o Promote exploration and characterisation efforts for *in-situ* populations determined to be at risk.
- o Assessing genetic diversity in collections with emphasis on:
- o identification of under-represented taxa
- o identify duplication among collections.
- o assess health and pest status of collections.
- o expanding and improving collection holdings.
- o Provide world catalogue information on citrus germplasm for incorporation in the global citrus information system.
- o Further development of conservation methodologies as alternatives to screenhouse/field collections.
- o Develop propagation technologies for citrus relatives.
- o Make avagation technologies for citrus relatives.
- o Make available general summary of established procedures for germplasm maintenance in implementation of certification programmes.
- o Promote implementation of certified nursery programmes in different citrus-producing countries.
- o Promote communications of recent advances in methodology for classical breeding, tissue culture, genetic engineering and gene isolation/cloning.

- o Promote collaboration on development of genetic improvement methodology.
- o Promote communication regarding field trials for rootstock and scion variety evaluation.
- o Promote collaboration on field evaluation programmes in order to generate more comprehensive information about citrus germplasm performance.

Global Network

The Network intends:

- o To facilitate inter-country and inter-regional scientific collaboration on tropical and subtropical fruit (TSTF) genetic resource conservation, evaluation and utilisation. The network is functioning on a voluntary basis involving the participation of interested national institutions as well as regional and inter-regional networks dealing with TSTF genetic resource conservation and utilisation.
- o To play a role in harmonising ongoing networking initiatives in different regions: UTFANET fooing networking initiatives in different regions: UTFANET for Asia and Pacific; RELAFRUT for the Americas; SEANUC for Africa; MESFIN for the Wider Mediterranean Region etc.
- o To link different initiatives in different parts of the world dealing with TSTF-including under-utilised fruits-genetic resource exploration, conservation and utilisation.

The specific tasks to be addressed by the network are the following:

Identification of genetic variability.

- o Identify primary and secondary centres of diversity for selected TSTF.
- o Facilitate cooperative efforts in genetic resource exploration.
- o Facilitate cooperation between PGR research work at the basic level and at field level.
- o Organise and develop documentation (*e.g.* catalogues).

Evaluation and Characterisation:

- o Facilitate cooperative studies on characterisation of TSTF through biotechnological or traditional means.
- o Revise criteria for existing descriptors, specifically to include utilisation criteria, as well as developing new ones.

Conservation.

- o To facilitate cooperative studies on appropriate and efficient conservation strategies for germplasm of selected tropical and subtropical fruits and their wild relatives.

Documentation and Information.

- o Facilitate the exchange of information among regional networks.
- o Diss
- o Facilitate the exchange of information among regional networks.
- o Disseminate information on collections, accessions, conservation technologies, catalogues, utilisation and plant genetic erosion concerning TSTF.
- o Facilitate development of a global database on TSTFGR.

Germplasm Exchange:

- o Help member networks and institutions on exchange of TSTF germplasm in full respect of all international agreements.

Utilisation:

- o Encourage regional collaboration to facilitate cooperative studies in TSTF genetic improvement.
- o Encourage the development of appropriate production technologies.

Transfer of Technology:

- o Identify the needs for technology dissemination in TSTF and promote the establishment of appropriate mechanisms to address them.

Public Awareness:

- o Promote efforts to increase public and institutional awareness about the importance of work on TSTF genetic resources.

With the Global Network providing a coordinating mechanism, the Mediterranean Fruit Inter-country Network (MESFIN), the RELAFRUT (with its Sub-network on Plant Genetic Resources Conservation), the UTFANET (Under-utilized Tropical Fruit Trees Network in Asia), and the SEANUC (South East Africa Network on Under-utilised Crops), as well as the new similar network being established in West Africa, will continue to be instrumental it work being established in West Africa, will continue to be instrumental in pursuing activities aimed at promoting the identification, evaluation and conservation of the existing genetic diversity of tropical and subtropical fruit species cultivated in the different regions, in order to make it available for utilisation in improvement programmes to increase and improve fruit production.

The International Network on Cactus Pear is carrying out a series of activities, including a survey of members' existing cactus genetic resource collections; the preparation of a "Descriptor list for cactus pear" and "Guidelines for improving and facilitating the exchange of cactus material for germplasm collections"; and the establishment of germplasm repositories in various locations. A basic germplasm information system for network members will also be established, following the criteria of the FAO World Information and Early Warning System on Plant Genetic Resources. The Global Olive Genetic Variability Network has recently been established and is undertaking activities related to all aspects of the identification, evaluation, and conservation of olive germplasm, and will assess the value of the world's available olive genetic variability, define priorities for its conservation, describe the facilities available in each country for conserving this germplasm, and elaboin each country for conserving this germplasm, and elaborate a mechanism for exchanging information through voluntary inter-country collaboration. The Network, which includes national institutions dealing with olives from all over the world, is preparing a world list of olive germplasm, covering both cultivars and wild relatives, and a description of their characteristics.

The Inter-regional Cooperative Network on Nuts with its transversal Working Group on Tree Nut Genetic Resources

Identification, Evaluation and Conservation assists its member countries in further developing a comprehensive programme of nut germplasm conservation, and interacts actively with their nut production sectors, in order to better guide a common and coordinated programme of nut genetic resource utilisation. The transversal working group is concentrating on preparation of nut germplasm catalogues, initially for walnut, almond and pistachio and subsequently for hazelnut, pecan, pine nut and chestnut. The Network of Traditional Crops for Southern African Countries (SEANUC) is establishing the state of genetic diversity of indigenous under-utilised vegetables, root and tuber crops, and also cereals and pseudocereals, grain legumes, in the region; and is recommending priority species for exploration, collection, conservation, evaluation and utilisation. A similar network is under preparation in West Africa.

In addition, FAO cooperates with:

The West Asia and North Africa Network (WANA); The Andean Network on Plant Genetic Resources (REDARFIT); The Amazonian Network on Plant Genetic Resources (TROPIGEN); and The Central American Network on Plant Genetic Resources (REMERFI).

Conservation Strategies

Modern plant breeders and bio-technologists rely on genetic variation in landraces, primitive cultivars and wild forms to produce better adapted and higher yielding crop varieties. Therefore, it is vital that a wider range of germplasm is conserved, both *ex-situ* and *in-situ*, so that it will be available in the future as a resource for adapting crops to new and changing environmental conditions and to sustain agricultural production and development.

With regard to horticultural crop germplasm, *in-situ* conservation strategies involving farmers more directly need to be considered, particularly for populations of wild crop relatives and wild food plants in the areas where they are found naturally. The term *in-situ* conservation also takes in the maintenance, by farmers, of domesticated landraces or traditional crop varieties. In the case of horticultural crops, home gardens in many parts of the world are examples of on-farm gere gardens in many parts of the world are examples of on-farm germplasm conservation efforts which deserve to be given greater support.

For horticultural species only, the tasks to be undertaken to achieve effective *in-situ* conservation of crop relatives and wild food plants are really too vast to be contemplated. Hence there is a need for setting priorities in terms of identifying those taxa and habitats for which protection is most urgently needed, and determining the most important, effective and efficient interventions to be undertaken.

Ex-situ conservation strategies, similarly, need to be based on rationalising criteria in order to reconcile the need, on one hand, to maintain collections of genotypes that include most of the traits needed for genetic improvement programmes with the necessity, on the other hand, to reduce gene bank maintenance costs and to minimise storage space requirements. In this regard, the concept of the core collection, *i.e.* the selected part of an entire collection which effectively represents the genetic diversity of a crop species and its wild relatives, fulfils a useful purpose for breeders and other users. Needless to say, the selection work entailed in establishing a core collection requires an intricate knowledge of the genetic variability of the taxa in question, and this can only be derived from exhan question, and this can only be derived from exhaustive studies on characterisation and evaluation of the contents of collections.

Biotechnology applications offer instruments that can add efficiency and flexibility to germplasm conservation systems in at least two different ways, *i.e.*:

- o by the use of molecular (biochemical and genetic) marker methods that enable the rapid characterisation of accessions; and
- o *in-vitro* conservation methods, providing space-saving and cost-saving alternatives for the conservation of material, in the form of slow-growth maintenance and cryopreservation techniques.

The *in-vitro* conservation methods are considered to be particularly interesting for those horticultural species that are normally vegetatively propagated as well as those with recalcitrant seeds, and continued efforts are needed in this field to develop protocols for a wider range of species.

Germplasm of many horticultural crops, particularly tree fruits, is usually conserved in field collections, due to the perennial nature of the plants. However, because of disease and insect problems, weather factors and improper management, many collections may be at risk, and germplasm may be lost. There is a need, above all, for systematic duplication of material for safety reasons, accompanied by information systems or safety reasons, accompanied by information systems that effectively tie together collections in different sites. From the strategy viewpoint, given that there are alternative approaches to the conservation of genetic variability of perennial crop species and their wild relatives, and considering the constraints of feasibility and cost associated with each, there is really a need to develop a system of complementarity between methods-field, seed, *in-vitro* and, when appropriate, *in-situ*.

Utilisation of Plant Genetic Resources

Evaluation and Characterisation of Germplasm

With regard to germplasm of horticultural crop species, many scientists agree that the lack of useful characterisation and evaluation data of accessions is often one of the main reasons for under-utilisation of the contents of genetic resource collections. In response to this problem, efforts are now being made for different horticultural crop groups to develop an approach to rationalising evaluation and characterisation work to fit within budgetary constraints and at the same time to take advantage of the possibilities offered by integrating the use of molecular tools. From such initiatives, notably within the framework of the different crop-related genetic resources networks, the following criteria are emerging and these may also guide collaborative initiatives to better the prospects f guide collaborative initiatives to better the prospects for the effective utilisation of genetic resources in crop improvement efforts for other species:

- o the development of a common set of criteria for the cataloguing and evaluation of material based on a rational combination of phenotypic characterisation and molecular marker methods.

- o the development of computerised information systems that should accommodate more complete and useful data entries, linking knowledge of traits for disease resistance, soil and climatic adaptability and horticultural performance, with genetic "fingerprinting" of accessions.
- o the implementation of common programmes, based on the outcome of the first two tasks, to examine and compare the accessions held in existing germplasm collections around the globe in order to determine the extent and usefulness of duplication and identify possible areas for cooperation in assessing the potential value of germplasm held and in rationalising collections so as to optimise resource use.
- o the development of a system to build up cost-effective germplasm conservation protocols that are based on an effective combination of field, *in-vitro* and, *in-situ* methods, supported with suitable duplication and documentation systems.

Plant Genetic Resource Development

Plant genetic resources serve as the basic raw materials for crop improvercces serve as the basic raw materials for crop improvement and together with seeds, they serve as the basis for food security. The conservation and utilisation of PGR is both an international imperative and concern at the same time. Following from that, the widespread utilisation of improved seeds by farmers is recognised as the most credible and cost effective strategy for enhancing food production.

It is therefore necessary to ensure that arrangements are in place to ensure adequate PGR management and to assure farmers of a continuous and unimpeded access to improved seeds and planting materials as a basis for food security. In the face of shrinking genetic variability as a basis for plant breeding and the supply of not more than 10 % of improved seed needs of farmers, urgent interventions are required in this area.

Much of the world's rural population is wholly dependent on its own farm-saved seed and planting materials for its food security. There is a need, recognised in the Global Plan of Action on PGRFA, to strengthen local capacity to produce, distribute and market farm-saved seed of crop varieties that are essential for local food security, and to help diversify agricultural production systems through the increased use and commercialisation of local and under-utilised crops.

The informal seed supply system has its roots in the age-old tradition used by farmers to ensure the supply of more than 90% of the planting mre the supply of more than 90% of the planting materials of staple food crops required to meet food security. This comprises mainly on-farm production, selection and saving of grain, tuber or stalk from harvested crop as seed for the next cropping season. It also includes on-farm seed production surpluses, which are traded in the appropriate local markets in various forms (cash, barter or kind).

The role of this system in germplasm conservation has been well documented. Among the many qualities which have made on-farm seed production a feasible option for seed security of staple food crops are:

- o Broader national coverage;
- o Maintenance and sustenance of many crops, which are not patronised by the formal seed supply system but which are vital to the survival of major sections of the population, particularly the under privileged;
- o Reasonable and affordable seed costs to other farmers and availability of alternate arrangements for seed payment;
- o Inclusion of farmer's varietal preference in the production programme;
- o Closeness of seed supply sources to cultivators; and
- o Stability of crop yield through the use of genetically broad-based varieties or ecotypes.

Thus, improved capacity for seed production is seen as a key to promoting the conservation and utilisation of PGR as a basis for food security and agricultural development, and for ensuring a better sharing of the benefits derived from ensuring a better sharing of the benefits derived from their use. FAO is assisting Member Countries in the development of their seed industries through the implementation of programmes which are consistent with their priority needs and in line with current international trends. The overall aim of such programmes is to stimulate the capability of the formal sector to provide farmers with their improved seed requirements in the most effective and efficient manner, whilst also assisting the informal sector, as well as individual farmer efforts, to make up the difference.

Recently, several programmes aimed to develop seed production for horticultural crops have been promoted, particularly in Africa, and through this regional exercise it has become very evident that research in this sector has made very little progress in the last 25 years. Most of the African countries are highly dependent on outside sources for horticultural seed and planting material. An inter-regional training programme for women seed producers launched in Central Africa, and the Improved On-farm Seed Production for SADC countries are programmes intended to help farmers produce good quality seeds and planting material from their own cultivars, and thus promote the conservation of their plant genetic resources.

In many parts of Africa today, improved seed usage through the formal sector is not more than 5%. It is critical that by the year 2010, than 5%. It is critical that by the year 2010, this should be raised to not less than 15 %. Since this target, though a big improvement on the present situation, falis short of overall requirement, the informal sector as represented by on-farm seed production will be expected to make up the shortfall. Hence, with regard to seeds and planting materials, the overriding requirement in seed industry development in Africa over the next ten years will be the strengthening of the informal sector, especially improvement in on-farm seed production, since this is the main source of seeds for small scale farmers.

Since the region is prone to natural calamities for which in the past emergency seed supplies have been a recurring feature, emphasis has been placed on the establishment of sub-regional and national seed security programmes involving a networking approach, to enable groups of countries to assist each other, using their own indigenous cultivars in meeting such emergency needs. The attainment of seed security will also involve the strengthening of national capabilities to ensure continuous availability of improved seeds. This entails assisting the countries of the region to develop appropriate policies and plans, to improve seed infrastructure and technology and to involve the private sector in seed production and distribution. To achieve this, emphasis is placed on training, including the training of women farmers in selected countriraining of women farmers in selected countries, and introduction of workable seed quality control schemes, such as the "Quality Declared Seed".

Plant Genetic Resources

Planned conservation and utilisation of local plant genetic resources is of prime importance for food security, as it permits small-scale farmers of developing countries to have permanent access to seed and planting material adapted to their region. Conservation of and access to these resources are most important in disaster-prone areas, as farmers of these regions often lose all of their seed stocks during disasters. Furthermore, disasters cause the loss of local varieties or genotypes, hence leading to the erosion of genetic diversity.

To ensure seed security, it is necessary to protect the vital genetic diversity of the important food crops and promote and strengthen the conservation of locally adapted varieties and genetic resources at the farm and local community level. Such conservation would complement, but not be an alternative to, national and regional gene banks. Though important to re-establish seed security quickly following disaster, locally conserved varieties are more vulnerable to disaster than nationally or regionally held genetic resources. Thus it is also necessary to develop early warning systems to monitor qualitative and quantitative changes in the status of geneticve and quantitative changes in the status of genetic diversity in locally and regionally adapted crop genetic resources.

A large amount of the plant genetic diversity that is lost in disaster prone regions also affects horticulture. During the rehabilitation process, after the event, it may sometimes appear easy to restore the vegetable production through the introduction of seeds of new adapted varieties. More difficult or even impossible may be to restore fruit production with appropriate varieties to be introduced from outside the region. In such circumstances, the choice of a wrong genotype may result in a second and irreversible disaster. Additionally, the introduction of such foreign varieties induces uncontrolled crossings with local genotypes adapted to the region. This "genetic pollution" is responsible for the definitive loss of local genetic resources and, as a result, of the disruption of the sustainable agricultural system of the region concerned.

From another point of view, horticultural staple food crops and horticultural traditional crops are certainly an important component of an effective seed security strategy as they may play a significant role in rehabilitating food production and income in a region affected by a man-made or natural calamity. Unfortunately, this is not always recognised.

It has therefore been considered appropriate to develop in this paper a chapter on this importato develop in this paper a chapter on this important topic, since the protection and preservation of crop genetic diversity should be addressed, as a first measure, in order to be able successively to develop strategies for the rehabilitation of agricultural production in regions affected by calamities through the seed multiplication and re-introduction of the same crops and varieties that were cultivated prior to the disaster.

Food Security

Seeds of the varieties and land races of our food crops provide the biological basis for plant food production, and seeds contain gene complexes which confer environmental and agro-climatic adaptation and resistance to diseases and pests. Plant genetic diversity thus makes a significant contribution to nutritional quality and food security.

The incidence of chronic undernutrition is unacceptably high in many poor countries, and without extraordinary efforts more than 20 of them could have still above 30% of their population undernourished by the year 2010. Even in countries with adequate or high average food availabilities, significant numbers and proportion of the population are and would remain inadequately nourished. And there is no country in the world without pockets of extremely poor, disabled, unattended for vulnerable individuals and groups, who cannot cater for lnerable individuals and groups, who cannot cater for their own needs.

As the outcome of the World Food Summit, FAO's Member Countries declared the objective that the incidence of undernutrition should be brought down to the incompressible minimum. This means there is an urgent need to achieve increased production in food deficit areas.

It is not possible to think that this increase of production can be achieved only by increasing the yield per hectare, with indiscriminate use of chemical fertilisers or pesticides. The possibilities of extending the areas used for agriculture are relatively limited and the yield capacity of some major staple food crops is reaching a plateau. The ecological and economic consequences of increased use of agro-chemicals and irrigation development are posing new threats to the sustainability of agriculture.

To reach the objective of increased food production, certainly it is necessary to explore opportunities for diversifying the crop base of agriculture, which in turn can be reached as an outcome of the sustainable exploitation of our plant genetic diversity. In this regard, horticultural crop development could indeed represent an important component of a global plan of action for improving food security, provided that coordinated research policies and adequate investments are addressed to improving the productivity of traditional and subsistence horticultural crops. Horticultural crosubsistence horticultural crops. Horticultural crops are only part of the total genetic diversity, nonetheless they can contribute greatly both to satisfying food security and to integrating and diversifying dietary needs of humankind. In fact, in many regions of the world, horticultural crops are important staple food crops.

In those developing countries where hunger prevails, primary importance is often given to the production of basic food crops, especially cereals, whereas the improvement of traditional horticultural subsistence crops-many of which are also rich in proteins, fats and carbohydrates-is too often neglected.

Political and public opinion have become more sensitive with regard to the conservation of genetic diversity for food and agriculture. However, not enough work is done to preserve crop genetic diversity and particularly horticultural genetic diversity. Hardly any specialised international institutions concentrate their efforts on preserving horticulture genetic diversity, whereas for basic food crops, important infrastructures and programmes have been developed in the last fifty years. In fact, only 16% of the accessions currently held in agricultural plant germplasm collections are of horticultural species, which is indicative of the vast amount of research work which still needs to be done.

It is also our current responsibility to promote awareness of the global problems of poveo promote awareness of the global problems of poverty, hunger and malnutrition, the long term solutions to which can be assisted by the committed efforts of the international community acting to ensure that governments, institutions, NGOs, industry and individuals work together as stewards of nature's diversity. Genetic resources, technology, information and funds are all essential and complementary resources in achieving food security and sustainable rural development. Using and sharing these resources in a fair and equitable manner is a moral obligation for the present generation and a condition for the survival of future generations.

Conclusions

Conservation and utilisation of plant genetic resources for food and agriculture are inextricably linked. Plant genetic resources should be made more easily available and useful to plant breeders and farmers for further improvement. The Global Plan of Action on PGR recommends a major initiative in evaluating existing collections. It recommends making the genetic material itself more readily utilisable through genetic enhancement and pre-breeding activities. Precautionary steps are needed in order to avoid genetic uniformity, which can increase the vulnerability of major crops. In both developed and developing countries, economic and social measures should encourage farmers who continue to grow local varieties and produce "dwho continue to grow local varieties and produce "diversity-rich" crops, on genes from which the development of new high-yielding crop varieties throughout the world depends. Hundreds of species used at local level (most often managed and harvested by women) are neglected, although many contribute substantially to household food security. These crops should therefore be given new attention, and marketing should be improved. The tangible added benefit of these and other initiatives would be the broadening of the genetic base of our food crops, thereby furthering crop stability and world food security.

As the Report on the State of the World's Plant Genetic Resources concludes, the better use of plant genetic diversity and resources will be a prerequisite to meeting the challenges of development, food security and poverty alleviation. In particular, greater use of plant genetic diversity and resources will be required in order to produce varieties adapted to the extreme and highly variable environments of low-productivity or marginal areas. In addition, the need to combine sustainable productivity increases with mounting pressure to reduce the use of agrochemicals, in parallel with the promotion of more efficient water and nutrient management, is likely to place even greater reliance on the utilisation of a wider range of PGRFA in high-productivity areas. To achieve the necessary productivity increases to keep pace with food demandvity increases to keep pace with food demand based on the sustainable use of natural resources, whilst maintaining the environment is a major challenge facing society. A more coordinated and concerted effort is needed to overcome all obstacles to the better use of plant genetic resources for food and agriculture to meet these objectives. This will require much strengthened collaboration between different countries, different scientific disciplines, private and public sector, NGOs and farmers.

The questions of the conditions of access to germplasm and intellectual property rights related to plant genetic resources are still being discussed in general terms within the framework of the international agreements and conventions cited in the previous chapters. Whilst mechanisms such as the global crop-related networks described above would not undertake to settle these issues, they could certainly constitute a qualified forum to advise governments and international bodies on the practical and longer-term implications of broader plant genetic resource policy decisions as they relate to the development of the horticulture sector.

Though this review exercise, it can be seen that a lot of progress has been made in developing policy related to the conservation of plant genetic resources, including horticultural genetic resources. However, it is not possible to state that this process is being back-stopped by adequate research po is being back-stopped by adequate research policy. In fact, if a global approach has been taken in different agricultural sectors where international agreements are already under negotiation such as the International Undertaking on Plant Genetic Resources, Farmers' Rights, Breeders' Rights, Plant Protection, etc., such a global approach is still lacking in the area of agricultural research, and in our case, in horticultural research. Indeed, the research sector is facing repeated drastic cut-backs in investment. Perhaps a lot of expectations are held for the applications of biotechnology as the means of resolving the problems of food and agricultural production in the third millennium. It is worth pondering, at this point, whether a more balanced distribution of investments among biotechnological research and classical research in the field, might produce more substantial results in terms of increased crop productivity.

Furthermore, the horticultural sector is perhaps today the most neglected vis-a-vis this globalisation approach. This is why FAO has very much welcomed the initiative to organise a World Conference on Horticultural Research and encourages the participants to use this forum for identifying and proposing concrete solutions and a globally coordinated plan of action for the development of horticultural crops for food and income.

It should be recognised that horticultural crops will play and recognised that horticultural crops will play a very important role in developing food production in the future. Most of the food crops traditionally cultivated in tropical and sub-tropical regions are horticultural crops and it is largely on these crops that efforts will need to be concentrated in order to meet increasing local food needs in many developing countries.

From a different point of view, if it is possible to state that progress has been made on developing global policies on broad thematic lines regarding plant genetic resources, it can be noted that through all this process of consultation, the specialised crop scientists and breeders seem somehow to have been disconnected from the decision-making fora and may find themselves facing decisions that have been made and ratified without their having contributed to them. Despite a lot of activity at the political level, initiatives to call all the stake-holders, and particularly the technicians involved in food and agricultural production, have not been so numerous. Consequently the influence of the scientific community in the policy-making process has been very limited, and colleagues would agree that it is time to correct this situation. In the field of horticulture, this World Conference on Horticultural Research seems to be the first initiative and this exercise should perhaps be continued and expanded at regional level.

Concerning the Conservatiegional Level

Concerning the conservation of horticultural plant diversity and its utilisation, it should be remarked that great efforts have been made in developing new cultivars to be addressed to clients of industrialised countries and in doing this, attention has been reserved to those traits that were highly adapted to the uniform management techniques of specialised modern agriculture. Efforts for preserving genetic diversity have likewise been targeted much more towards commercial end uses, rather than to protecting other aspects of diversity which could be critical in adapting crops to more marginal agro-ecological conditions. In other words, very little has been done so far for the poor agricultural regions in developing countries.

We have remarked that food production will need to be increased by about 60% in the next thirty years, and most of this increase should occur in developing countries. It should therefore be recognised that research programmes should also include activities on topics addressed to improving the productivity of those traditional crops that may not be considered remunerative from the economic viewpoint of industrialised countries but which represent a sustainable source of food for populations in poor agricultural regions. What could be suggested, as an innovative approach is a co-sharing of responsibilities between industrialised and developing countries in jointly defining and implement countries in jointly defining and implementing research policy and programmes in order that development assistance is not only expressed in terms of financial support, but also through effective cooperation among specialised institutions and scientists, with the institution of an associated mechanism for monitoring progress.

Agro-forestry in Shifting Cultivation Control Programmes in India

About 7000 B.C., according to the archaeological evidence (Childe 1956) mankind began to change its mode of life from food gathering to food cultivation, by adopting shifting cultivation which is, historically, a transitional stage towards more permanent agriculture. As the years passed, the peoples living in river valleys advanced rapidly and brought about a number of institutional and technological changes; but meanwhile some of the people living in remote areas continued in the blind alley of shifting cultivation and failed to take part in the progress towards culture and civilization. However, in the development of a country all sections of society need to be involved, including the backward shifting cultivators, all are equally entitled to the benefits of civilization and development. For these reasons programmes for the control of shifting cultivation have become important, particularly in tropical countries, and agro-forestry has played a pivotal role in these programmes.

Shifting Cultivation Practice

In general, hill people who live in tropical regions practice shifting cultivation. After a piece of land is selected, trees or bushes are cut down partially or fully, left to dry and then burnt in situ. In the cleared land seeds of crops are dibbled into holes or broadcast, without using ploughs or animal power. When the crop yields begin to decrease after some years the cultivator moves to a new patch of forests to repeat the process, and allows the abandoned land to recuperate. After a period varying from two to twenty years, he returns to the same land for cultivation. This practice is also known as shag. "slash and burn agriculture", swidden, jhum, etc. The system aims at self sufficiency and does not lead to any capital formation or growth. Human labour and land, in its widest sense, are the chief factors of production. It is both a labour-intensive and land-extensive system, based on low technology.

There are two divergent views about shifting cultivation one condemning it and another, a liberal one, upholding it as a humane practice. The first one, often termed an "outsider's view", states that it dries up the springs of the hills, causes soil erosion, destroys valuable forests and adversely affects rainfall and deprives people of benefits of forest produce. The second one, often called an "insider's view", considers it as "an organic response of the people engaged in it to certain specific ecological conditions, rather than to a particular ecotechno system... It is crude but it is interlaced with the way of life of people who possess a crude technology and very little capital" (Bhowmik 1976). The opinion of M. D. Chaturvedi, former Inspector General of Forests (Chaturvedi and Uppal 1953), may be taken as a representative statement:

The notion widely held that shifting cultivation is responsible in the main for large-scale soil erosion needs to be effectively dispelled. The correct approach... lies in accepting it not as a necessary evil, but recognizing it as a way of life; not condemning as an evil practice, but regarding it as an agricultural practice evolved as a reflex to the physiographical character of land.

Shifting cultivation is mainly confined to medium to high rainfall zones, between 100 metres and 2,000 metres above sea level. It is practiced in three types of vegetation, forests, bush and grasslands.

The density of population in the area under shifting cultivation varies from three persons per km^2 in Arunachal

Pradesh to over 300 persons in Tripura or Drissa, but people mainly depend on shifting cultivation when the population is below thirty persons per km^2. The study of the land/man ratio is important in working out shifting cultivation control programmes, as at a high land/man ratio, shifting cultivation does not become a problem.

The lengths of the period under cultivation and of the following fallow period vary considerably from place to place and according to the community involved and depend on population density and local environmental factors. When the increase in population upsets the ecological balance, the people cultivate the area for longer periods, reduce the fallow period, encroach upon unusable land, search for new jobs, reduce consumption pattern or migrate to other areas.

The cultivation pattern is not uniform at all places. The Noctes and Nissis of Arunachal Pradesh Pollard Macaranga denticulata trees at the time of clearing the forests. When they vacate the area they sow seeds of these trees in blank areas so that they will be available to provide house-building material and firewood when the people return. The Wanchos leave Schima wallichii trees standing in the area. In Nagaland, alder (Alnus nepalensis) trees are pollarded and left in the area. In Madbya Pradesh, the Abujhmarhs leave mohwa (Bassia latifolia). In Andhra Pradesh, toddy palm (Borassus flabellifer), tamarind (Tamarindus indica) and mango (Mangifera indica), all of which provide food, are left by the hill people while clearing the land for shifting cultivation.

The cropping pattern also varies from place to place. In high rainfall areas, rice is grown as the principal crop, with mixed cropping of maize, millet, cotton, etc., as a secondary crop. In medium rainfall areas, oil seeds and maize are grown as the first crop, followed by millets and rice. In low rainfall areas, millets are sown along with vegetables and pulses.

This traditional system of cultivation was well adapted to the environment, and the people's knowledge of growing cereals and tree crops enabled them to maintain an ecological balance. However when the population increased they were unable to find ways of raising the long-term carrying capacity of forest lands. In addition, the population increase in the lowlands and the effects of industrialization have caused increased demand for saleable produce, and the lure of money has caused the hill peoples to increase the area under shifting cultivation. Thus the problem has become one of unprecedented dimensions, demanding an immediate solution.

The pattern of land ownership is important in the success of programmes for the control of shifting cultivation. In the north-eastern states, where no permanent land settlement has taken place, shifting cultivation is practiced in the three following categories of land ownership: (a) lands owned by villagers collectively; (b) lands owned by the chiefs who distribute lands among the individual households for shifting cultivation; (c) lands owned by individual families. In this region decisions for development can be taken quickly and this has, to a large measure, been responsible for the success obtained.

In the other states shifting cultivation is practiced either on unsurveyed areas, or as encroachments on land belonging to the government. This uncertain legal status of the land has been an impediment to the speedy and timely organization of development programmes.

Some data are available on shifting cultivation and are given below, but are far from complete.

Experiments conducted by the Indian Council of Agricultural Research, Shillong (Borthakur 1976) have shown that in hilly areas, the bottom third of the slope can be used for agricultural crops on terraces, irrigated from low dams; the middle third can be used for fruit trees and cash crops such as coffee, black pepper, big cardamom,' etc., planted along the contours; while the upper third should be used for forestry and fodder crops. This could form a basis for the planning of the land-use pattern in programmes for shifting cultivation control.

A study of 920 sq. km in Arunachal Pradesh showed that the value of forest produce burnt annually in the area under shifting cultivation would be about U.S.$600,000.

Soil analysis studies in shifting cultivation areas in Arunachal Pradesh and Tripura made by Jha et al. (1976) indicated that shifting cultivation caused an appreciable change in organic matter content resulting in nutrient imbalances, and reduction in water-holding capacity, sesquioxides, iron, aluminium, nitrogen, calcium, magnesium, potassium, phosphorous, cation exchange capacity and C: N ratio. The pH value increased, *i.e.* the soil became more acid.

Agro-forestry

Agro-forestry has been defined as a "sustainable and management system which increases the overall yield of the land, combines the production of crops (including tree crops) and forest plants and/or animals simultaneously or sequentially, on the same unit of land, and applies management practices that are compatible with the cultural practices of the local population" (King and Chandler 1978). Though agro-forestry is not new, during recent years its importance has increased dramatically especially as regards its potential for optimizing land use in the tropics. Its primary aims are the production of food and wood, and conservation and rehabilitation of soil resources needed for future production, at the same time maintaining and improving the quality of the producing environment. In the coming years growing population combined with increasing pressures on finite areas of agricultural land will make the food supply situation even more precarious. There will not only be more people on the land to feed but there will also be an increasing demand for food and wood for the growing urban populations.

According to estimates made by the United Nations Environment Programme, up to 5,000 km^2 of land are lost to productive use every month in the world because of erosion, flood damage, salinization and alkalization, advancing deserts and other causes. In developing countries, potential arable land is not adequate to meet the growing needs of the increasing population. In India, nearly half of the land area is subjected to water and wind erosion. Since irrigation facilities are limited, the choice for the future is limited with relatively small opportunities to bring new arable land under the plough. This calls for a new approach and in this respect agro-forestry could help to produce food and wood while conserving the ecosystem.

Sometimes it is said that agro-forestry is suitable only for marginal and brittle ecosystems. In reality it can be practiced on all types of agricultural lands, as it enables better utilization of the nutrients and water available in the soil as well as of solar energy.

Agro-forestry can be effectively practiced on lands subjected to shifting cultivation, on mountain ecosystems denuded of vegetation from biotic causes and in arid and semi-arid tracts. In agro-forestry two essential and related aims are (a) the conservation and improvement of the site, and (b) the simultaneous optimization of the combined production of a forest crop and an agricultural crop.

Through proper selection of tree species it should be possible to minimize soil erosion, to tap nutrients from deeper levels than those reached by the roots of agricultural crops, and to replace through leaf fall and fixation of atmospheric nitrogen the nutrients removed in the crop. According to E.F. Schumacher (1974) "a most marvellous 3-dimensional incredibly efficient contrivance already exists, more wonderful than anything man can make, the tree. Agriculture collects solar energy two-dimensional; silviculture collects it 3-dimensionally. "

Much research remains to be done on the choice of species, the relationship between trees and other crop plants, and the development of multiple-use trees (trees which can be used for fodder, or fruit, or other produce in addition to wood production); this includes research into genetics and breeding.

Sometimes it is said that for successful implementation of agro-forestry programmes, land hunger and unemployment should exist among the people with a low standard of living. However, success of this system depends on the incentives given, social amenities and services provided and marketing facilities arranged. So, like any other rural development scheme, an agro-forestry programme has to be considered with relevance to the social and economic development of the people. For successful implementation of agro-forestry, the institutional requirements are also important. The curricula and syllabus in training and educational institutions need to be properly oriented by making changes. What is essential is that scientists and institutions should be "people oriented" and they should develop systems that are appropriate to the physical, biological and socio-economic conditions that prevail in the country.

Agro-forestry in Shifting Cultivation Control Programmes

Past Work: Since shifting cultivation forms part and parcel of the day-today life of tribal people in India, the policies adopted by government towards tribal people have also affected shifting cultivation practices. In the pre-independence era, two distinct approaches were adopted towards the tribal people. These were total isolation by segregation of tribal areas and de-tribalization, leading to assimilation by others. The first approach created pockets of tribal society leading their lives in their own style and manner while the second made them third-rate citizens on the lowest rung of society. After independence, the policy adopted was given in the following words of Jawaharlal Nehru: "We do not mean to interfere with their way of life, but want to help them to live it according to their own genius and tradition" (Verrier 1959).

The Jhum land Regulation 1948 of Assam was not helpful as it covered only limited areas of northeast India and was difficult to enforce. The National Forest Policy, 1952, which laid stress on weaning people away from shifting cultivation by persuasion was unable to achieve any notable success.

Efforts to regulate, control and contain shifting cultivation have been made in several states from time to time with varying results.

The taungya system, though it succeeded in West Bengal and Kerala and resulted in the creation of forest villages which provided labour for forestry operations, was not successful in other states as it was unable to mobilize all the shifting cultivators and was confined to certain groups, particularly landless and lowland migrants. The introduction of intensive methods of permanent cultivation was also unsuccessful as it failed to cater to the needs of all the shifting cultivators and was not suited to the socioeconomic life of the tribal people. Since 1970, with the emphasis on micro-level planning, the tribal sub-plan area programmes were introduced in selected places. These catered for the special needs of the tribal people with an integrated area development approach and had special funds. These programmes, by and large, were successful. Other programmes were also taken up by the state governments and the central government, like the "Pilot Project on Control of Shifting Cultivation" in eleven states in 1976-1977 and the Small Marginal Farmers Development Agency during 1975-1980.

The North-Eastern Council, Shillong, took up programmes in the north-eastern states for control of shifting cultivation. These were mainly aimed at soil and water conservation by building terraces for development of agriculture and the production of horticultural crops including tree crops. They covered an area of 11,360 hectares at a cost of U.S.$6 million.

The most remarkable feature of shifting cultivation is that almost all the varieties of cereals and vegetables, together with tree crops, are grown in the hills in a single field. This is impossible on wet plain lands. This "cafeteria system of cultivation" yields the cultivator better returns in kilogrammes of grain per man-hour of labour input than does settled agriculture. In efforts to replace this system by a better method, agro-forestry is useful as it is linked to the existing system.

Himachal Pradesh, one of the Himalayan hill states, in about six years from 1970, was able to achieve remarkable success in hill area development by growing apples as a cash crop. It evolved a three dimensional forestry combining silviculture and horticulture with animal husbandry.

In the northeast region, certain tribes of Arunachal Pradesh and Nagaland consider the possession of mithuns as a symbol of prestige and status. Most families in Mizoram and Nagaland rear pigs. The number of cattle, pigs and poultry per hundred humans is larger in the northeast region than the average of the country, as a whole, which indicates the importance of animals in the economy of this region.

The experience gained indicates that wherever an integrated area development approach involving agriculture and forestry was made, success was achieved. Animal husbandry could supplement the income of the people. Thus agro-forestry has come to be the mainstay of the shifting cultivation control programmes. The experience gained in the various states is described in the following paragraphs.

Andhra Pradesh: 116,000 people, Konda Reddi, Samantha, and others, practice shifting cultivation on 8,392 km^2 in the agency area, in the northeast corner of the state, mostly on unsurveyed forest, but also to some extent on areas encroached from government forest. They generally leave useful trees such as Caryota palm, mango, tamarind, toddy palm (Borassus flabellifed and jack fruit ((Artocarpus mtegrifolia) in their cultivation area, to provide them with edible products. To take advantage of this, seedlings of these species have been supplied to them. Also, in suitable areas, besides rice and millet, cash crops such as coffee under the shade of silver oak (Grevillea robustal and jack fruit, tobacco, oil-seeds, and turmeric are grown, the produce being marketed through Girijian Cooperative Society. The centrally sponsored scheme "Pilot Project on Control of Shifting Cultivation," was taken up in 1977 and has been partially successful.

Assam: Shifting cultivation is practiced on 4,980 km^2 of unsurveyed land in two hill districts; the people are autonomous in character. Among the 403,000 tribal people the Karbis, besides cultivating maize, vegetables, potatoes, cotton, ginger, chillies, etc., also grow castor, tapioca, and mulberry (Mows laevigata) on which to rear silkworms. In the shifting cultivation control programme the main emphasis was in providing areas for cultivation of wet rice, and settling the people there. Shortage of flat land prevented much progress being achieved. The "Pilot Project on Control of Shifting Cultivation" was taken up in 1977 and has been partially successful.

Bihar: 61,000 tribal people, Paharias, practice shifting cultivation on private land, and to a limited extent, on land encroached from government forest, in Santhal Parganas district. They grow rice, millets, pulses and maize, and often rear tassar silkworms on sam (Terminalia tomentosa). The programme aimed at settling them on flat lands, for wet rice cultivation, has been successful where such land was available.

Madhya Pradesh: 14,000 tribal people, who are Hill Maria, practice shifting cultivation on about 3,900 km^2 on unsurveyed forest land in the interior of Bastar district, though this is not officially encouraged.

They generally grow rice, millets, beans, pulses, maize, sweet potatoes, chillies, oil seeds and vegetables. They protect, and often plant, the following trees: mohwa (Bassia latifolia), pales (Butea monosperma) samul (Bombax ceiba), kirni (Mimusops indica), ber (Zizyphus jujube), munga (Moringa oleifera), palms (Caryota urens, Phoenix sylvestris, P. farinifera), and siari (Baubinia vahlii). Attempts to resettle them in the valleys have only been partially successful.

Manipur: About 300,000 people practice shifting cultivation on about 1,000 km; of unsurveyed land in the hilly areas. They grow rice, maize, potatoes, yams, chillies, vegetables, millets, mustard, sugarcane, sesame, ginger, turmeric, cotton, etc., and take care to preserve trees of commercial value in their cultivation areas. Attempts to establish permanent cultivation in the valleys have only been partially successful, due to lack of suitable land.

Meghalaya: About 350,000 people practice shifting cultivation on about 4,160 km^2 of unsurveyed land. Their crops include, depending on altitude, potatoes, rice, maize, millet, yams, sweet potatoes, tapioca, oil seeds, cotton, etc. They also cultivate areca nut (Areca catechu), betel leaves (Piper belle), bay leaves, black pepper, pineapples, citrus fruits, bananas, etc., and in some places rear eri silkworms on mulberry, tapioca, and castor. Permanent cultivation on terraces in the valleys has been developed, but its extension under the shifting cultivation control programme has been limited by lack of land. Coffee, under the shade of jack fruit and silver oak, has also been raised in places.

Mizoram: About 260,000 people practice shifting cultivation on about 6,040 km^2 in the hilly areas. They grow rice, maize, millet, sugar cane, beans, chillies, ginger, turmeric, yams, tobacco, etc. They also rear tassar silkworms on a large scale on oak (Quercus lamellosa and Q. serrate). Lack of land has limited the development of permanent terraced cultivation in the valleys, and the introduction of plantation crops such as coffee and tea has been to a limited extent only.

Nagaland: 400,000 people practice shifting cultivation over an area of about 6,080 km². Rice is the dominant crop, but millets, potatoes, yams, chillies, cotton and maize are also cultivated. They allow alder (Alnus nepalensis) to come up in their fields; besides providing small timber, charcoal, firewood, etc., it also helps to increase the nitrogen content of the soil as it has nodules containing symbiotic bacteria. They also grow the leguminous tree Parkia roxburghii for its edible beans, and raise tassar silkworms on oak in places. Development of terraced agriculture in valley bottoms has been done only to a limited extent. There is a proposal to grow coffee over extensive areas under the shade of silver oak and jack fruit.

Orissa: 706,000 people practice shifting cultivation on about 37,000 km² in hilly areas in the northern, central and southern parts of the state, partly on unsurveyed land, and partly on land encroached from government forest. Besides rice, millets, pulses, vegetables, turmeric, tobacco, maize and oil seeds are grown, with jack fruit, bananas and pineapples among them. There has been only partial success in establishing permanent cultivation in the valleys and on the lower slopes. The planting of coffee under silver oak has been taken up over extensive areas, and it is proposed to extend this programme.

Tripura: 100,000 people practice shifting cultivation on 2,208 km², in the southern and eastern parts of the state. These areas are unsurveyed, though records have been kept of them at the village level. Besides rice, both early and late maturing, they grow millets, beans, pulses, cotton, oil seed, sugar cane and jute, with among these crops bananas, jack fruit, citrus and pineapple.

Permanent terraced cultivation has enabled people to settle in the river valleys. A shifting cultivation control programme in the hills, based on an integrated multi-disciplinary approach, has had fruitful results. Selected families have been allowed to develop orchards, reclaim land for raising field crops, and improve lakes and water areas for fisheries. They have been given housing assistance, advice on animal husbandry, primary education, and medical and other facilities over five years. In addition each family has, for seven years, been allowed to raise

1 hectare of rubber plantation under the technical guidance of the Forest Department; they have also been paid wages. This agroforestry programme has been quite successful, and there is a proposal to take it up over extensive areas (Ghosh 1982).

ArunachalPradesh: 270,000 local people practice shifting cultivation over 2,486 km^2, on hill slopes throughout the state. Besides rice, both early and late maturing, they grow millets, maize, cotton, pulses, yams, chillies, tobacco, mustard and sugar cane, with pineapples, bananas, orange and jack fruit among them. Permanent cultivation on terraces in the valleys and on the lower slopes has been successful wherever irrigation has been provided.

Shifting cultivation control programmes have been undertaken successfully in two fields, one by devc, ping areas already under shifting cultivation, and the second in a programme of retaining forests as an area for future expansion of cultivation (Thangam 1979).

In the areas under shifting cultivation an integrated area development programme was used. In general this consisted of terracing the lower third of the slopes for agricultural crops, with or without irrigation; growing horticultural and other cash crops on the middle third; and using the upper third for forestry and fodder crops and grasses. The crops to be grown were selected after a proper land-use survey, designed to get the maximum return from the land. The people were encouraged to raise pigs, and to rear fish in ponds formed by impounding water in small water courses. Wherever possible, Low-level earth bunds were constructed and small diversion channels built to irrigate terraced areas for permanent cultivation.

Big cardamom (Amomum subulatum) was found to grow well in the area, beginning to yield from the fourth year onwards. During the first four years this cultivation provided employment for at least one person per family; from the fifth year onwards the crop yielded about U.S.$300 per hectare, rising to $600 from the tenth year onwards. Each family could plant two or three acres of this crop, and so could obtain an annual income of about $1,000, including income from other crops.

Various other development work was undertaken, in soil conservation, road construction and afforestation, among others. This was done after consultation with local village leaders, who fixed the priorities for the work. Other facilities such as health, education, etc., were provided, with the help of the local government.

The work began in the Mintong village in Tirap district in 1978, and its success has induced three neighbouring villages to offer their land for similar development programmes.

The other programme was to keep forest areas as a reserve for future expansion. In 1977, with the help of local village councils, such areas were formed into village reserved forests, and managed on a scientific basis to obtain the maximum benefit from the forests. The net profit obtained was shared equally between the Forestry Department and the village council, who utilized their share to create durable community assets. At Tafragaon, Lohit District, the management of 98 km^2 of such forests made it possible to start a residential school for fifty Mishmi girls, at a cost of U.S.$50,000.

The adoption of agro-forestry principles has contributed to the success of both programmes.

Future Programmes

It has been seen that the integrated area development approach, in which agro-forestry played a leading rote, has contributed to the success of programmes for shifting cultivation control. Tribal people practicing shifting cultivation raise agricultural crops and trees simultaneously, as they need the former for food, and the latter to provide timber and bamboos for house construction, agricultural implements, fencing, etc., and also to provide a good burn when the land is opened for agriculture. They see no dichotomy between agriculture and forestry, and only wish to obtain an economically beneficial way of life from their land. The study of their methods will help to evolve better ways for the development of multiple cropping and inter-cropping, and of agroforestry.

Methods to be used in programmes for shifting cultivation control will not be uniform for all states or for all tribal people within any state. Tribal customs differ, as do the varieties of crops which will grow in different localities. The somewhat backward Wanchoo people in Arunachal Pradesh willingly devoted their land for cultivation of big cardamoms, while in Tripura the rubber plantation programme succeeded. The object should be to produce the best return from the land, with the help of the people living there, as far as this is consistent with environmental factors.

In programmes for shifting cultivation control, unless the yield from the land was maximized through the adoption of agro-forestry methods suitable for the area, and unless the willing cooperation of the local people was obtained, unqualified success was not achieved. Programmes where plantation crops such as coffee, large cardamoms, and medicinal plants such as Dioscorea, combined with agro-forestry, brought in quick returns, were more acceptable to the people, and in addition helped to conserve environmental factors and maximize production.

In selection of areas for integrated area development programmes it is often said that such areas should be compact and easily manageable. It has been recommended that entire watersheds or catchments should be brought under treatment. This may be technically sound, but in practice catchment areas often cover several villages, not all of which may agree to the proposed programme. So it has been necessary to set up programmes for single villages, developing the micro-catchments within the village area, and ensuring development of this area as a whole. Such small and compact areas will not only be easy to handle, but will also enable results to be achieved and shown in the field in a short time.

In these programmes emphasis has been on the development of locally available resources in manpower and natural resources with the minimum investment. No import of technology is required. The use of such intermediate technology in agro-forestry programmes has been successful in the past. The low level of productivity has been improved through maximizing work opportunities to enable the people to live a better life. This will not only eliminate underemployment and unemployment, but will also help to stem the migration to urban areas. This form of development requires a capital investment of only about U.S.$1,200 per hectare, and this amount can be obtained from financial institutions on loan. In Arunachal Pradesh, a ten-year project on the above lines, in which 100,000 families will each create a plantation of a least 1 hectare of cash crops such as coffee, large cardamoms, or medicinal plants such as Dioscorea, will shortly be begun after the necessary finance has been provided by the Agricultural Refinance and Development Corporation of India, who have agreed to the proposal. Under this scheme the annual income per family will rise by U.S.$300 after the fifth year, and this income will double by the tenth year. The income of these families will be raised above the poverty line, and so Arunachal Pradesh will be the first state in India to raise the living standards of nearly 90 per cent of its inhabitants above the poverty line in such a short period.

Shifting cultivation practices are linked with the ecological, socio-economic, and cultural life of the people and are closely connected to their rituals and festivals. The theme of food production, that is shifting cultivation, is interwoven through the whole gamut of life of primitive society, whose whole philosophy of life is a product of the system of economy of people who possess only crude technology and very little capital. This is the reason why many new methods of cultivation recently introduced into tribal societies have failed to win cultural acceptance.

The rituals and festivals of the people are organized to take place at various stages of the shifting cultivation year. Unless the people are convinced of the advantages of new agricultural methods, they will not voluntarily abandon shifting cultivation. But once they are convinced, they will change their practices, and will then also be able to adjust their ways of living, their rituals and their customs in a suitable way.

In Arunachal Pradesh the Apatanis (Furer-Haimendorf 1946) and the Monpas have more or less settled down to permanent agriculture in permanent fields, as a result of various circumstances. However this has proved no impediment to the continuation of their rituals and festivals, while their economic status has considerably improved. Our aim should be to achieve such a happy combination, through a proper psychological approach. Thus the voluntary involvement of the people in a programme, and its acceptance by them, are a sine qua non for success.

The absorption of technology is not merely a matter of calculating costs and benefits, prices of technology, costs of labour displaced, or revenue received for the product. Technology has an impact on society, the mode of living and the relationship between people and institutions. Technological change cannot be abruptly imposed. It has to be a process of evolution. However modern, however beneficial a technology, it has to fit in with indigenous culture and capabilities and harmoniously transform tradition into modernity. Erik P. Eckholm (1976) has said:

It is generally easy to recommend technological answers to ecological problems. Political and cultural factors are invariably the real bottlenecks holding up progress. Changing the relationship of man to land in the mountains, as any where else, invariably involves sensitive changes in the relationship of man to man.

Conclusion

In programmes for shifting cultivation control, agro-forestry, which aims at the production of the best return from the land through integrated area development programmes, has been successful. With the help of institutional finance, it is possible to adopt agro-forestry programmes over a span of time and with the willing participation of shifting cultivators, help them to improve their socio-economic status, and thus enable them to join the mainstream of the more advanced people of the country.

Citrus Fruits Basic

Citrus (*Citrus* spp) is grown in almost all the states of India. Although the citrus industry in India has faced many challenges, there has been a consistent increase in area and production owing to the awareness for sustained production. Citrus fruits, consisting of mandarin orange (*Citrus reticulata* Blanco), sweet orange [C. *sinensis* (L.) Osbeck] and lime [C. *aurantifolia* Swing le] are grown commercially in tropical, subtropical, arid irrigated and mountainous regions in varying soil and weather conditions.

Citrus is grown practically all over India. However the states of Andhra Pradesh, and Maharashtra have the largest share. Although citrus trees on the whole do well in dry climate, with a rainfall between 75 and 125 cm, certain species, such as pummel and certain mandarin oranges, thrive in heavy-rainfall areas of Konkan, Assam and Coorg.

Citrus trees the grown in almost all kinds of soil, varying from heavy black soils to shallow open soils. Some of the varieties of citrus seem to adapt themselves to soil conditions better than others. They thrive in free-draining alluvial or medium black soils of loamy texture. A hard substratum or a sticky impervious layer is very injurious. Soils having a high water-table should be avoided. Click on individual citrus fruits to view their basics:

Grapefruit

The name grapefruit (*Citrus paradisi* Maef.) has been derived from the habit of bearing the fruit in clusters like grapes. In India, its introduction is comparatively recent, and its cultivation is confined mostly to Punjab, the western parts of Uttar Pradesh and to places around Poona in Maharashtra.

Climate and Soil: The climate and soil requirements of the grapefruit are similar to those of the orange. High rainfall and humidity are harmful, as they encourage diseases.

Varieties: The popular varieties, which are all imported, are 'Marsh Seedless', pink-fleshed 'Foster' and yellow-fleshed 'Duncan,

Propagation and Planting: Propagation is done by budding. However, owing to their polyembryonic nature, seedling-trees have been frequently found to be quite satisfactory. The rootstock most successfully employed in the northern regions is kharna khatta (Citrus karna Raf.). In the south and Bombay-Deccan, 'Jamburi' is commonly employed, whereas in Assam grapefruit does well on Rabab tenga. Planting, irrigation, manuring and interculture are the same as for the orange.

Pruning: Grapefruit trees require less pruning than orange-trees.

Harvesting and Marketing: The harvesting season is from January to March in the north and from September to November in the south. Picking, packing and other operations are the same as for the orange. Quality and flavour of the fruit is improved if it is stored before transporting.

Lemon/Lime

(Citrus limon (L.) Burm. f) is not cultivated to any great extent in India, as it requires a comparatively cool climate for regular bearing. Its fruit is not so highly flavoured as that of sour lime. The lime (Citrus aurantiifolia Swingle), both sour and sweet, known as kaghzi nimboo and mitha respectively, are more prized in India than lemon. Sour lime is propagated mainly from seed. Budding on rough lemon rootstock, layering and moreotting are also practiced to some extent. The tree is susceptible to frost. It flowers twice a year in February-March and again in August. The main crop is obtained in August from the first flowering. The second crop is ready in the following February. The propagation of sweet lime is done from mature wood cuttings, which root readily. It can also be propagated from seeds and the seedlings, usually come true to type. Planting of limes and their pruning, manuring, etc. are the same as for the sweet orange.

Mandarian Orange

Area of Cultivation: Mandarin is largely grown in Arunachal Pradesh, Karnataka, Maharashtra, Meghalaya, Nagaland, Punjab, Tamil Nadu, Tripura and West Bengal. In Punjab, Kinnow is a commercial cultivar, whereas ecotype of Nagpur mandarin is grown in different regions known by different names..

Climate and Soil: Santra or mandarin orange (Citrus reticulata Balanco) grows successfully in all tropical and subtropical parts of the country. It tolerates more humidity in summer and winter than the sweet orange. It is grown under rain-fed conditions in Coorg, Wynad tract, Palni Hills and the Nilgiris in the south between elevations of 600 and 1,500 metres. In Assam, the main centres of production are the Khasi, Jaintia and Lushai Hills. The region around Nagpur (elevation about

370 metres) produces a superior quality of mandarins. It is mainly grown under irrigation. In Punjab, its cultivation is confined mainly to the submontane districts up to about 600 metres. It can be grown successfully on a wide range of soils, but the ideal soil is medium or light loam with slightly heavier subsoil. Heavy black soil, underlain with murram and having good drainage, is also suitable. In the Khasi Hills of Assam, oranges are grown on sandy or gravelly soils.

Varieties: The important varieties cultivated on a commercial scale are the 'Nagpur' orange, the 'Khasi' orange, the 'Coorg' orange, 'Desi Emperor' and the 'Sikkim orange.

Propagation: The propagation of mandarin orange is largely through seed, except the 'Nagpur' and 'Emperor' varieties, which are propagted by budding. Like other citrus species, the seed is polyembryonic. Therefore, while propagating by seed, the sexual seedlings, which are usually stunted, and poor are rogued out and the rest that are produced from the cells of the nucellus are allowed to grow. The seedlings, thus selected, are more or less uniform in growth and production. They are, however, late in bearing and remain tall and slender. Budded plants do not suffer from these defects. The santra orange is usually budded on rough lemon (jambhiri, Soh-myndong or jatli khatti) rootstock. The variety 'Emperor' is budded on the kharna khatta rootstock. Studies at the Citrus Experiment Station, Coorg, of the Indian Institute of Horticultural Research, has shown that Trifoliate, Rangpur lime, Kodakithuli and Troyer citrange are good rootstocks for mandarin.

Planting: In the hills and humid regions, where plantings are generally done on steep slopes, the land is properly terraced. In the plains, where the trees have to be irrigated, the land should be leveled. The trees are usually transplanted during the monsoon. In heavy-rainfall areas, the planting is generally done at the end of the heavy rains. They are planted 4.5 to 6 metres apart.

Pruning Prune young trees to build up a strong framework, as recommended for sweet orange. The bearing trees require little or no pruning. Undesirable growths, like water-shoots and crossing branches, should be removed once or twice a year. In

Bombay-Deacon, root exposure or resting treatment is given to santra trees to make them flower to order. The treatment is the same as for the sweet orange.

Manuring: Farmyard manure, 20 to 25 kg per tree, is applied at planting, together with about half a kilo of each of N, P and K per tree may be applied in the first year after planting, and the dose is gradually increased to 0.45 kg of each of N and P and to 0.90 kg of K per tree in the seventh year and kept constant thereafter. The dose of farmyard manure is increased to 50 kg per tree. It may be replaced by green manuring. In northern India, manuring is generally done in winter, whereas in Bombay-Deccan it is done before the advent of the monsoon or at the time of root exposure.

Water Management: Citrus plants responds well to irrigation. Water management constitutes an important input in citrus orchard, which determines productivity and longevity. Normally no irrigation is practiced in heavy rainfall areas of North-Eastern states. Plants that receive 1 or 2 irrigation during dry period have healthy growth and produce heavier fruits in these areas. At the same time consecutive irrigation without definite period of soil-moisture stress is detrimental to crop yield, as such as situation leads to more vegetative growth. Monitoring of soil moisture is effective for regulating flowering and fruiting in acid lime under tropical situation. Of late drip irrigation is becoming very popular in water-scarce area of Maharashtra, parts of Tamil Nadu and Andhra Pradesh. Many young citrus orchards were planned and planted wit drip system of irrigation. The economic analysis of these 2 systems favoured for the use of drip irrigation, which is viable. The system improved the yield, uniformity in fruits and gave brilliant quality of fruits in respect of shape, colour and edible quality. The growers apply the nitrogenous fertilizer through the drip which further economize on the fertilizer use.

Harvesting: Seedling trees bear their, first crop in the eighth year and the full crop from the tenth year onwards. Budded trees start bearing from the fourth year and full crop is had from the seventh year onwards. The harvesting periods differ in different parts of the country. While picking the fruit, the stem-end should be cut close to the fruit without damaging rind. Packing is done by putting the fruits of different size grades in separate wooden crates.

Sweet Orange

Area of Cultivation: In sweet oranges, Musambi is largely grown in Maharashtra, whereas Sathgudi or Chini is the basis of commercial cultivars in Andhra Pradesh. Blood Red Malta, Malta, Jaffa and Valencia are grown in Punjab.

Season: Dry and arid conditions, coupled with distinct summer and winter having low rainfall, are most favourable to the growth of the sweet orange. Rainfall seems to be unimportant if irrigation is provided, but atmospheric humidity exerts a great influence.

Soil: The sweet orange can be grown on a wide range of soils, from heavy clays to very light sands, with pH ranging, from 6.0 to 8.0. The tree is particularly sensitive to high concentrations of salts and cannot stand water-logging.

Varieties: The important varieties of sweet orange grown in each, region are 'Blood Red', 'Pineapple', 'Hamlin', 'Jaffa' and 'Valeneia Late' in northern India, mosambi in Western India and 'Sathgudi' and 'Batavian' orange in southern India.

Propagation and Planting: Usually, propagated by budding. The most suitable rootstock is Jamberi or jatti khatti. For 'Pineapple', 'Hamlin', 'Jaffa' and 'Valencia Late' varieties, kharna khatta also provide a suitable root-stock. The trees are planted 6 to 7.5 metres apart each way in January or August-September in the north and in July-August in the south. The bud-union should be kept at least 15 cm above the ground while planting.

Pruning: The pruning of citrus-trees begins in the nursery. All branches that start within a few centimetres of the union are removed, leaving about half a metre of clean straight stem with a few well-placed branches. All unwanted branches are removed once a month during the first year after planting, and once in two to three months in subsequent years. The bearing trees require little or no pruning. After the crop is picked, the branches touching the ground should be cut close to the laterals so that no stubs are formed. All diseased injured and crossing branches, water-sprouts and dead wood should be removed periodically.

Root Exposure: In the Bombay-Decean region root exposure is given to the year. Water is withheld for about two months in advance of the normal flowering season, and after about a month, the roots are exposed by removing about 10 cm of the soil in the case of light soils and about 20 cm in the case of heavy soils. After about 10 days, the soil is returned mixed with manure, and a light irrigation is given. After four or five days, a more copious watering is given, followed by 10 days later by the full dose of water. In the case of light soils, the withholding of water without root exposure is sufficient to check vegetative growth and force blossoming. In southern India, no root exposure is given nor is it feasible. In the north, the root-exposure treatment is not necessary as the trees normally rest in winter and flower once a year. It should be mentioned that in most situations, the root exposure of citrus trees is a devitalizing process and should be resorted to only under expert advice and direction.

Manuring: Manuring may be followed as in the case of the santra orange.

Banana Technical

Banana (*Musa* sp) is considered to be the fourth most important food crop in world after rice, wheat and milk products, based on gross value,. Banana is being cultivated in India from antiquity, It has a great socio – economic significance and is closely interwoven in our national heritage. Banana is a rich source of starch, mineral and vitamins and thus makes very good food for babies. Banana is also useful in managing patients with high blood pressure and heart diseases as it contains low amount of sodium, very little fat and no cholesterol,. It is also good for arthritis, kidney diseases, ulcer and gastro-enteritis. In India banana is grown widely in different agro-climatic conditions having different production systems and cultivars. India is credited to be major centre of diversity of Musa clones. At the same time, research efforts have resulted in substantial improvement in production and productivity of banana. Large number of cultivars totaling about 70 are grown in different parts of the country, depending on production system and regional preference. Many cultivars have poor yield potential but are grown for its quality and farmers are compensated with higher price of the produce. However, Dwarf Cavendish and Robusta known by different names are basis of commercial cultivation owing to high yield, wide market acceptability, short-crop duration and high economic returns per unit area.

But these cultivars are not common in coastal region due to their high susceptibility to sigatoka leaf spot disease. Poovan is another cultivar grown commercially in different regions for its wider adaptability, tolerance to drought and diseases. Nyali Poovan (Elakki Bale or Ney Poovan or Safed velchi) is also grown commercially in many parts of the country. Rasthali is significant in commercial production and its success largely depends on higher price it fetches. Virupakshi, Monthan, Karpooravallli and Chakia are also important in some regions. Nendran or Rajeli is grown in Kerala and Tamil Nadu. The clonal situation prevailing in different regions suggests that varietal need cannot be unified and would need regional consideration. Investigations have also enabled to identify the cultivars for different situations to maximize the returns. Grand Naine, an introduction is becoming popular in some parts of the country.

Varietal Improvement

According to the Department of Agricultural Research and Education, ICAR, Government of India, in 1999-2000 diagnostic norms and interpretation techniques-DRIS and CND-were developed to improve diagnostic precision for Robusta and Ney Poovan bananas. Yield limiting factors and overall balance of nutrients were identified via Nutrient Imbalance Index (NII) which helps in rational fertilization. The optimum N, P and K in leaf of Robusta were 2.2-3.5, 0.09-1.17 and 2.66-5.11% respectively. In Ney Poovan, it was 1.42-3.56, 0.06-0.20 and 1.44-3.69% N, P and K respectively. A wide variability was observed in M. balbisiana and ornamental species found in the deep forests of Arunachal Pradesh. Four different balbisiana clones, 2 each of diploid (BB) and triploid (BBB) genomes were noted with unique characters. Two wild accessions were also collected from shilling. One unique wild accession which is non-suckering and seed forming, the only means of propagation, was collected from hills of Diphu in Assam. The seeds were large (8-10 mm long and 5-6 mm across) with a smooth seed coat. Two hybrids H 1 and H 2 were released as BRS 1 and BRS 2 for cultivation in Kerala. Plant spacing of 1.8m x 1.8m for Poovan and Kothia, 1.2m x 1.2m for Rajapuri and Basrai, 1.2m x 1.5m x 2.4m for Gandevi Selection and 1.2m x 1.2m x 2.0m for Palayankodan was cost-effective. Planting 3 suckers at a spacing of 1.8m x 3.6m with 300:90:450g/pit of N: P_2O_5: K_2O respectively gave highest cost: benefit ratio. Tissue-cultured plants performed better with higher yield with reduced crop duration and uniformity in fruits.

Crop Production

Owing to the adoption of improved production technologies which include season of planting, selection of planting material, method of planting, high density planting, adoption of drip irrigation, nutrition and integrated management of pests and diseases, tremendous increase in production and productivity in banana has been observed in the past 50 years.

Propagation

Seed-sterile nature of the commercial cultivars and vegetative parthonocarpic has made it almost mandatory to use vegetative part for propagation. The sword suckers of 700-1,000 g are optimum. Rhizomes with active lateral buds and dead central buds are preferred for distant transportation in western India. Double paring and shade drying followed by dipping in 0.5% monocrotophos and 0.2% Bavastin is recommended to disinfect against nematode and soil-borne fungi. In-vitro propagation of banana has been achieved successfully and this method is becoming a commercial reality. However, the adoption of in-vitro propagated plants among growers has been slow mainly because of the high costs. However, adoption of drip irrigation and better management system, has resulted in enhanced return from in-vitro plants. Uniformity in flowering and early harvest of in-vitro plants have been demonstrated at different locations.

Weed Management

Weeds cause approximately 60-70% yield loss in banana. Vegetative phase (1-6 months) is most critical and weeds during this phase of growth cause a considerable loss in productivity. Control of weeds during this period enhances the fertilizer-use efficiency and the yield. Suppression of weeds using intercrop, and control through integrated management results in increased production. Growing of crop cowpea [Vigna unguiculata (L.) Walp.] is most effective in suppressing the weed growth. Spray of glyphosate twice proved equally effective.

Biofertilizers

Azospirillum enhances the height, pseudostem girth, and leaf production and leaf area of Poovan banana. In Elakki Cale, association of mycorrhyza increases yield. Under the present system of cultivation, yield of banana exhibits plateau in commercial plantation, and to increase yield exogenous application of growth-regulators have to play significant role. Among the several bio-regulators tried at different locations, application of 2, 4-D (10-20 PPM) improved the bunch weight. Bunch covering with polythene in November increased yield by 25-30%. Two sprays of KH_2PO_4 at fruit-development stage also increased the bunch weight in Basrai banana.

Pest Management

Banana weevil (Cosmopolites sordidus) is of major concern in Kerala, Andhra Pradesh and Assam among the insect pests. According to experts, chemical control for this pest has been worked out. Though mites (Tetranychus sp.) and thrips also cause considerable damage to flowers and fruits, the situation is not alarming. Banana aphid (Pentalonia nigronervosa) is wide spread and causes the damage through transmission of banana bunchy top (BBT). In north-eastern regions, Bihar and Bengal, problem of scarring beetle is alarming which causes scarring on the fruits peel by feeding on young flowers and fruits. The fruits affected by this pest have poor market acceptability.

Severe incidence in growth phase reduces yield. Pouring of 30 ml of 0.05% monocrotophos in whorl has been found effective. Pseudostem-borer (Odoiporus longicollis) affects the tall cultivars, which can be controlled by application of systemic insecticides and sanitation of orchard. Other insects are of minor importance. Although problems of insects and pests vary from region to region and chemical control measures have been worked out, enhanced research efforts, directed towards the investigation on factors causing appearance and development of population are needed to develop new techniques with emphasis on integrated management.

Nematodes

Numerous nematode species parasitize on banana roots and their importance as a problem in banana production largely depends on production system, region and cultivars. Although presence of burrowing nematode (Radopholus simiis), root lesion (Pratylenchus coffeae), root-knot nematode (Meloidogyne incognita), spiral nematode (Helicotylenchulus multicinutus), (Helicotylenchulus dihystera), cyst nematode (Heterodera oryziocola) are reported from different parts of the country. Burrowing nematode appears to be of wide spread. Helicotylenchulus is more serious in north-eastern region, whereas cyst nematode is a problem in Kerala only. In Maharashtra where banana is grown in rotation with sorghum problem of nematode is less significant. Similarly, in wet land system of production nematode is not serious compared with garden land banana. Double paring and treating the sucker with 0.5% monocrotophos, growing of sunnhemp as an intercrop and application of neem-cake invariably reduces the population of nematode. Chemical treatment is also suggested. But integration of available cultural, biological, chemical and genetic possibilities is better answer, which would need the knowledge on plant-parasite-soil interactions.

Disease Management

Considerable loss in banana production is caused by fungal diseases like sigatoka leaf-spot (Mycosphaeralla musicola) in humid tropics or coastal regions. Tolerance to the disease depends on the cultivar and growing conditions. The regions where this disease is virulent susceptible cultivars should not be grown, as fungicides are not very effective in reducing incidence of the disease. Monitoring of disease through forecasting, cultural practices, use of fungicides and genetically manipulation needs better understanding, as the disease behaviour is influenced by production system and prevailing weather conditions. Fusarium wilt (Fusarium oxyporum f. sp cubense) is serious on Rasthali group of banana. Fortunately, Canvendish group of banana is not susceptible to this disease, although sporadic incidences has been recorded. Flooding the soil, selecting the suckers and injecting the Carbendazim are effective measures in reducing the incidence. Solution to the problem is possible only through the genetic manipulation.

Interaction of fusarium wilt with nematode also needs increased research efforts. Erwinia soft rot and disease caused by Collectotrichum gleosporium and Anthracnose are of regional importance. Large banana plantations have been destroyed by banana bunchy top virus (BBTV). It is considered to be one of the more serious problem. Collection of plants from disease-free area, timely rouging and sanitation are effective in checking the incidence. Aphid vector, *viz.* Pentalonia nigronervosa is present in all the regions and even the virus-free material does not remain free for long time due to secondary infection. Tetrazolium test has been found effective in detecting the virus, but its sensitivity to little change in chemical makes the test unreliable. Antioxidants are used to enhance the reliability. Quick detection using monoclonal antibody was found effective and can be used for monitoring disease.

The losses caused due to this disease largely depend on the stage of infection. In-vitro plants appear more susceptible to this disease. If the spread of disease is not checked, it can be one of the major threats. Use of disease-free planting material, avoidance of susceptible hosts and control of vectors (Pentalonia nigronervoga) are some of the approaches for the management of the disease. Use of antibody has been found effective in monitoring the diseases. Banana streak virus (BSV) identified last year is found to be present in all the banana-growing regions on Poovan cultivar. The symptom is characterized by conspicuous chlorotic leaf streak, which becomes progressively necrotic. The symptom expression is highly influenced by weather conditions and becomes severe in ratoon crops. The pathogen is transmitted by the mealy bug (Planococcus citri) in a semi-persistant manner.

The disease is cause of concern, as it is spreading to other cultivars also. In-vitro plants have more incidence of this disease. Kokkan disease of Nendran banana reported from Kerala in 1966 continued to be riddle, a disease of unknown etiology, which was identified as banana bract mosaic virus (BBMV). The characteristic symptoms of the disease are spindle-shaped pattern in unusually red coloured pseudostem, dark streak on petiole base, reddish streaks on bracts and undersized unfilled fruits with an ultimate loss of yield. This disease is widely prevalent on monthan group of banana in all the southern states. Use of disease free planting material, rouguing of infested plants and sanitation are recommended for the management of this disease.

Other Latest Developments in Banana Research

The National Research Centre on Banana (NRCB) at Tiruchi in Tamil Nadu, established by ICAR in 1993, has a 36 hectares farm located 14 km southwest of Tiruchi. Besides carrying out research and development work on improvement in production technology, plant protection, post-harvest management and genetic resource management, the Centre is also providing consultancy services for developing banana orchards. The Centre maintains a banana germplasm of 597 accessions with collections made through the International Network for Improvement of Banana Plantain based in the Philippines. Tiruchi is a banana district in Tamil Nadu with largest area under banana planted with almost all cultivars under different systems of cultivation. The Centre for Plant Protection Studies (CPPS) of the TNAU, Coimbatore and the NRCB at Tiruchi have developed a technique to identify the banana punch top virus at an earlier stage under a project taken up with financial assistance, the World Bank. The scientists at the regional research stations of All India Coordinated Research Project on Banana and the NRCB at

Tiruchi are trying to develop hybrid varieties of banana. They have already developed two hybrids, named FHIA-01 and FHIA-03, which were on trial in 1998.

Papaya Technical

Papaya (*Carica papaya* L.), is a table fruit in most of the tropical countries cultivated mainly for fresh fruit consumption. It is consumed as a part of breakfast dessert and as fruit salad. It is also used in soft drinks, jams, ice-cream flavouring, crystallised fruit and also sold as canned cubes and juice in some countries. In India it is mostly consumed as a table fruit. Papaya research during the last 5 decades encompassed both basic and applied areas. Research work concentrating mainly on crop-production technology, areas suitable for different agro-climatic regions of the country had been carried out at different agricultural universities and ICAR institutions. Systematic papaya breeding work for development of improved varieties was initiated at the TNAU and the IARI in early sixties and subsequently at the GBPUAT and the IIHR in the seventies.

The first improved cultivar of papaya Coorg Honey Dew was identified in 1959 at the Citrus Experiment Station, Chethalli. Interdisciplinary research programme on crop improvement, production technology and plant protection is presently under implementation at the centres at Coimbatore, Pusa, Pune, Jalgaon, Rahuri and Bhubaneshwar with the support under the All-India Co-ordinated Research Project on Tropical Fruits of the ICAR, and through ad-hoc projects under ICAR AP Cess Fund Scheme.

Varieties: 15 high-yielding varieties of papaya have been developed in India, out of which 9 are widely adopted. Some successful hybrids have also been developed and heterosis has also been noticed with a wide range of breeding materials. Some of the common varieties in India include Coorg Honey Dew, Pusa Dwarf, Pusa Giant, Pusa Majesty, Pusa Delicious, CO.1, CO.2, CO 3, CO.4, CO.5 and CO.6. Varietal improvement work for frost tolerance, disease resistance and improved fruit quality is under progress.

***Varietal Improvement*:** According to the Department of Agricultural Research and Education, ICAR, Government of India, in 1999-2000, the Pink Flesh Sweet variety recorded maximum fruit weight, whereas Red Indian maximum TSS (13.0° Brix). Hybrid Surya having medium-sized fruits (600-800 g), red flesh, high TSS (14° Brix) and low cavity (20%) with good keeping quality was identified to be released. The Indian Institute of Horticultural Research, Bangalore and developed a hybrid variety of papaya, 'Surya', which is a cross between 'Sunrise Solo' and 'Pink Flesh Sweet'. It produces medium size fruits of 600-800 gram with red pulp about 3.0-3.5 cm thick and a smaller fruit cavity. The hybrid is gynodioecious and does not require male plants for pollination. The seedlings for planting. If planted at a spacing of '6x6' may yield 48-58 tonnes papaya per acre.

Propagation: Papaya is usually propagated by seeds. Germination of seeds takes place 2-3 weeks after sowing. Seeds treated with thiourea (100-200 ppm) and gibberellic acid (GA3 at 200 ppm) germinated better. Seed germination is highest at an optimum temperature of 35°C. Storage of seeds at 10°C was found to be the best for retention of seed viability. Viability of papaya seed can be maintained for 9 months of room temperature when stored in air-tight containers. Patch budding has been reported to be quite successful (90%) in papaya but it is not practiced commercially. Tissue-culture technique has been standardized for papaya micro-propagation. It has been reported that plants raised through tissue-culture technique yielded more fruit in Pusa Delicious when compared with the seedlings. Young papaya seedlings form callus profusely in Murashige and Skoog medium supplemented with auxins (NAA).

Callusing was obtained from stem segments, roots and leaf tissue. Callus readily differentiated into leafy shoots in medium containing high ratio of kinetin and NAA (2.0: 0.2 mg/litre), while root formation was favoured by a low kinetin: NAA ratio (0.2: 0.5 ml/litre). Shoot tip culture from seedlings and lateral buds from female plants of Coorg Honey Dew variety was found to be successful. The tissue-culture plantlets measuring 5.5 to 9.5 cm and hardened in plastic chamber survived better when planted in the field. The available information indicates that tissue-cultured papaya plantlets can be used with added advantage of uniformity in plant type.

Plant Density: A spacing of 1.8 m x 1.8 m is normally followed in most of the places. Field trials conducted indicates that a spacing of 1.8 m x 1.8 m for varieties like CO. 1, CO.2 and Solo, and 2.1 m x 1.2 m (3.968 plants/ha) for Coorg Honey Dew variety was optimum for Bangalore conditions. The spacing of 1.4 m x 1.4 m or 1.6 m x 1.4 m is ideal for CV Pusa Delicious under sub-tropical conditions of Bihar. For the dwarf variety Pusa Nanha a closer spacing of 1.25 m x 1.25 m has been found satisfactory. For papain production a spacing of 1.6 m x 1.6 m was recommended based on field trials conducted in Tamil Nadu. In West Bengal under a spacing of 1.5 m x 1.5 m (4, 444 plants/ha) fruit yield of 98.7 tonnes/ha could be recorded with papaya variety Ranchi.

Nutrition: The CO. 1 papaya plant revealed significant uptake of nutrients like N, P, K, Ca and Mg at the flowering and the nutrients were removed by the whole plant at harvest were 305, 103, 524, 327 and 183 kg of N, P, K, Ca and Mg/ha respectively. Thus the ratio of N:P:K removed was 1:0.33:1.71. The lack of N and P causes reduction in growth of the papaya plant, while absence of K and Ca induces pronounced foliar symptoms. Among the micro-nutrients, B followed by Fe, Cu, Zn and Mo are considered important for proper growth and development. Deficiency of B and Zn in field conditions causes considerable yield loss. The NPK requirements of papaya have been determined by conducting large number of fertilizer trials at different regions of the country. Field trials showed that 140-350 g N, 200-300 g P_2O_5 and 200-300 g K_2O/plant/year are optimum, depending on the cultivars and soil-climatic conditions. The applied nutrients to leaf showed optimum concentration of NPK in leaf petiole as 1.52% nitrogen, 0.14% phosphorus and 4.42% potash. Organic manures like farmyard manure (FYM), sheep manure, wood ash and bone-meal improved yield and quality of fruits of papaya. Inorganic fertilizers @ 200-250 g each of N, P2O5 and K_2O/plant are generally recommended for obtaining high yield in papaya. The N @ 200 g/plant was found optimum for fruit yield at Pusa (Bihar), but papain yield increased with doses up to 300 g/plant at Coimbatore (Tamil Nadu). Some of the important fertilizer recommendations emerged from field trials conducted at different parts of the country are as follows:

i. A dose of 250 g N, 250 g P_2O_5 and 500 g K_2O/plant/year applied in 6 bimonthly intervals was found the best for Coorg Honey Dew variety and gave estimated fruit yield of 186.8 tonnes/ha in 28 months crop cycle at a plant spacing of 2.4 m x 2.4 m (1, 736 plants/ha).
ii. A dose of 350 g N, 250 g P_2O_5 and 200 g K_2O/plant year applied in 6 split doses was best for Solo variety spaced at 2 m x 2 m (2,500 plants/ha) under Bangalore (Karnataka) conditions.
iii. At Coimbatore, a dose of 200 g each of N, P_2O_5 and K_2O/ plant in 4 split doses during 1, 3, 5 and 7 months after planting resulted in higher yield of papaya variety CO. 1. For papain production from CO.2 variety a dose of 250 g N/plant/year in 6 split doses at biomonthly intervals commencing from the second month after planting was found the best.
iv. A fertilizer dose of 200 g N, 300 g P_2O_5 and 600 g K_2O/ plant gave highest fruit (110 tonnes/ha) yield in papaya variety Ranchi under West Bengal conditions. For maximum papain yield 94.45 g/fruit), fertilizer dose of 200 g N, 300 g P_2O_5 and 400 g K_2O was found optimum.

Application of phosphatic fertilizers reduces acidity, while potash normally increases the TSS and total sugar. When K was adequate, TSS did not lower even with high N fertilization. The fruit quality in papaya as indicated by TSS was significantly influenced by K and N x K interactions. Not many field trials have been conducted to determine the micronutrient requirements of papaya.

The B-deficient plants normally develop bunchy top and deformed fruits. Foliar applications of Mn (0.25%) plus Cu (0.25%) at 60 and 90 days after transplanting helped in harvesting maximum fruit yield (94.7 tonnes/ha) with high quality fruits. Combined application of B (0.1%), Mn and Cu markedly improved fruit quality in papaya variety Ranchi under West Bengal conditions. Beneficial effects of mycorihiza have been observed in papaya. Association of vesicular arbuscular mycorrhiza (Glomus mosseae + Glomus fasciculatum + Gigaspore margarita) helped in saving 25% of P recommended for papaya.

Water Management: Papaya is susceptible to waterlogging and the orchard should have good drainage system, particularly in heavy soils experiencing high rainfall. Total water requirement of CO.2 variety of papaya is estimated to be of 1,800-1,900 mm and irrigation at 60-80% available soil moisture (ASM) depletion was found optimum for papaya. Plants receiving frequent irrigation with 100% evaporation replenishment maintained higher transpiration rate, lower diffusive resistance but gave significantly higher yields as compared to 50 and 25% evaporation replenishment. Evapotranspiration decreased with decreasing frequency of irrigation. Generally irrigation to grown-up plants at 7-10-day interval during hot summer months is recommended. Among various systems of irrigation, ring of furrow system of irrigation is preferred over the bed system. Bed system allows direct contact of irrigation water with the plant and thus predisposes plants to collar rot disease. Irrigation in widely spaced furrows of 100 cm resulted in significantly higher fruit yield as compared to basin and flood methods of irrigation under Bangalore condition. Drip system of irrigation helps in saving 50-60% water, and water use efficiency under drip is much higher.

Weed Management: Chemical weedicides like atrazine and simazine were found phytotoxic for papaya. Gramoxone at 1.0 litre/ha controlled weeds and gave maximum fruit yield in papaya (91 tonnes/ha) under West Bengal condition. Paraquat (1.0 litre/ha) applied twice at 90 and 180 days after transplanting caused maximum reduction of weeds, improved plant growth, advanced flowering and resulted in higher fruit yield. In field trials conducted at IIHR, Bangalore, Diuron 2 kg ai was found effective in controlling weeds without any adverse effect on papaya.

Disease Management: Various diseases, caused significant loss in yield and quality of papaya fruits every year. Some of the more common diseases and their recommended remedies are given below:

Nematode Diseases

Application of 1 kg of neem-cake or Carbofuran @ 2 kg/ha. Certain other minor fungal diseases like Phytophthora; fruit rot, stem rot caused by Rhizoctonia solani; anthracnose caused by Colletotrichum gloeosporioides, leaf spot caused by Alternaria sp and Cercospora sp; and leaf blight caused by Helminthosporium rostratum have been reported in the country. Certain post-harvest diseases also cause great losses during transportation and storage of fruits. As most of the causal organisms are associated with fruits from the pre-harvest states, application of protective fungicides is most effective in reducing the losses.

It is advised to frequently spray fungicides like Mancozel or Chlorothalonil at pre-harvest stages. Hot-water immersion of fruits after harvest at 42°C for 30 min, followed by a 20 min immersion at 49°C reduces the inoculum level substantially. The hot water-treated fruits receiving fungicide wax emulsion treatment are free from post-harvest decay and store well. A hot water (54°C for 3 min) spray treatment is developed for effective control of decay caused by anthracnose and stem-end rots. Breeding work involving resistant species is in progress. Through repeated selection and back-crossing the population has reached to F7 generation and some degree of resistance to PRSV is also reported. The breeding work at Pune since 1968 has resulted in identifying a resistant line having small fruits (0.5 to 1.2 kg) with TSS ranging from 7 to 10°Brix. At IIHR, Bangalore, hybrid progenies from crosses of C. papaya x C. cauliflora are tolerant to PRSV. Post-monsoon planting of papaya reduces the incidence of the disease under subtropical condition of north India. A ready-to-serve papaya juice drink has been developed at TNAU, Coimbatore. Papaya candy or tutti-fruitti is being produced at commercial scale in Coimbatore, Bangalore and Jalgaon.

The product is in great demand in bakery and confectionery industry. According to DARE, Solo papaya recorded good growth and higher accumulation of boron. Fruit set increased by 20% either by soil application or foliar spraying. Application of boron increased Ca in fruits which is an important element to increase the shelf-life of papaya. Papaya plants receiving inorganic nutrients alone (200 g each of N, P2O5 K2O/plant/year) recorded highest yield, but inclusion of organic source of nutrient (farmyard manure/neem cake) produced better quality fruits. Drenching with Ridomil (0.2%) was most effective method to control root-rot.

Post Harvest Management of Papaya

Papaya fruits are sensitive to poor quality outturns and high post-harvest losses if harvesting, treatments and handling techniques are inadequate or inappropriate. From harvest, a shelf-life of four to six days under tropical conditions and up to three weeks at low temperature storage can be achieved with the correct harvest maturity, disease control measures, handling techniques and storage conditions.

Harvest Maturity

Papaya fruits should be harvested when the colour of the skin changes from dark green to light green and when one yellow streak begins development from the base upwards. Fruits in this condition will continue to ripen normally after harvest. Those fruits harvested before this stage will fail to show complete ripening, and those harvested after, are more susceptible to damage and bruising during handling.

Harvesting

Papaya is harvested manually depending on the size and age of the tree, using specialised tools, knives or by hand. When harvesting by hand or with knives, the peduncle is snapped or cut from next to the tree.. then immediately trimmed flush against the top of the fruit. The specialised implement for harvesting of fruit inaccessible by hand due to tree height, comprises of a long pole, a small circular hoop at the top, a small mesh bag attached to the hoop, and a horizontal blade above the hoop and the bag. The blade is positioned below the peduncle of the fruit and the pole moved upwards; the fruit is detached from the tree and then drops gently into the mesh bag below the hoop at the top of the pole.

After harvest, the fruit are placed in single layers into shallow, light coloured field crates, preferably containing a foam layer for cushioning. All stems should be trimmed after harvest to ensure that no stem to fruit rubbing occurs during transport to the packing facilities. Fruit should never be thrown or dropped. Field crates containing the fruit should be left in shaded conditions protected from the sun and rain, while awaiting collection for delivery to a packing facility. Mesh bags, sacks or baskets are unacceptable for papaya transport due to the high susceptibility to bruising. Care should be taken during transport in field crates to minimise the movement of fruit.

Export Grading and Packing

Grading should be carried out as soon as possible after harvest, and fruit left under ambient conditions to continue ripening or placed at 10° to 12°C for cooling and storage. On arrival in the packing facility, fruit should be washed in water to remove latex and debris, then treated in a 0.05% Thaibendazole solution for anthracnose control. Washing, treatment and grading can be carried out using mechanised or manual systems, depending on the volumes of fruits. Grading in each carton is required in terms of size, sex (shape) and stage of ripeness. Female and hermaphrodite fruit cannot be mixed in the same carton; all fruits must be of a similar size in each carton resulting in a range of counts, and separations must be made for the degree of ripeness. Carton net weight is dependent on the importer, ranging from 3.5 to 5 kg (8 to 11 lbs), and must not be overfilled during packing.

Packaging

Single layer, one-piece self-locking or two-piece full telescopic fibre board carton; bursting strength 200 to 250 lb/in2. Internal packaging material should include shredded paper in the base of the carton and individual tissue wraps for each or alternate fruit. Individual labels can be attached to the fruit for appearance and recognition.

Carton internal dimensions:

- 10.9 by 34 by 26.9 cm (4.3" by 13.4" by 10.6")
- 10.2 by 43.2 by 27.9 cm (4" by 17" by 11")

Storage and Transportation

Importers require fruit at specific stages of ripeness for optimum sales; this varies between 50 and 70% yellow colour depending on the importer, the market and the time of year. For fruit to arrive in the importing country at the correct colour stage, attention has to be paid to the maximum and minimum colour stages on departure from the packhouse, the length of the shipment period and the temperature in the importing country.

Thus, colour stages on departure from the packhouse for air shipment can vary from 20% to 50%, and the selections are generally more rigid during the summer months due to the rapid rates of ripening on arrival in importing countries, fruit exported too green (less than 20% yellow colour) will fail to ripen adequately when temperatures in importing countries are low, particularly in winter months. Sea-shipment of papaya is possible when fruits are shipped at the optimum harvest maturity, with one or two yellow streaks. Shipments should be made at 10° to 12°C and 85 to 95% relative humidity, in refrigerated holds or reefer containers. Post-harvest disease control is critical with sea-shipment, particularly anthracnose and Phytophthora. Colour development during sea-shipment usually increases from 10% to 40% during 10 days at 12°C, and will develop further during the customs, clearance and delivery period on arrival.

Fruit harvested and placed to ripen at the recommended harvest stage (one yellow stripe) will ripen to 60 to 70% yellow colouration within four to six days under ambient tropical conditions (25° to 28°C). Fruit transferred to low temperature storage (10° to 12°C), when harvested at the one-stripe stage, will store successfully for 14 to 21 days if post-harvest disease incidence can be controlled. When harvested at more advanced stages of ripening, the storage life will be significantly reduced.

Potential Post-harvest Losses

Papaya is particularly susceptible to post-harvest losses as a result of high susceptibility to bruising and disease infection. Careful handling must be employed during harvesting, handling and shipping, and the relevant disease control measures employed. Mechanical Damage: Damage to the skin immediately after harvest, as a result of the harvesting implement, dropping into crates, over-filling of crates and excess movement of fruit during in-field transport, will result in latex staining, punctures, scars and bruises.

During ripening, bruised areas will develop into dark soft regions, which become affected by secondary microbial infection. Similar effects can occur as a result of poor handling during washing, grading and packing. Damage can be reduced by taking protective measures throughout the handling procedures. Staff should be trained with harvesting techniques, foam should be included in the base of field crates and crates should contain only one layer of fruit. Stems are to be removed in the field to prevent puncturing or scratching of adjacent fruit. Vehicles used to transport the fruit from the field to packhouse should be driven slowly and with care. During handling in the packhouse, fruits should never be thrown or dropped and in automated operations, all machinery should be padded where possible. Palletised systems of produce movement in packhouse operations are preferable, both with field crates and with the final packed product. Pathological Factors: Anthracnose (Colletotrichum gloespoirides L.) disease is prevalent during long term storage and with humid orchard conditions, inappropriate pre-harvest disease control or poor orchard hygiene.

The presence of the disease is characterised by small black or light brown spots, which gradually enlarge and may coalesce and sink. Anthracnose can be controlled or reduced by pre-harvest sprays. Post-harvest control where pre-harvest sprays are in operation should include a cold-water dip or spray containing 0.05% Thaibendazole and surfactant. Changing of chemical is required each four to five tonnes of fruit. Where no pre-harvest control is in operation and the disease is present in the orchard, the cold water treatment may have to be replaced with a hot-water Thaibendazole treatment. A treatment time of 20 minutes at 46°C is required. Specialised equipment for temperature maintenance and water circulation is necessary as fluctuations in temperature will reduce the effectiveness of the treatment and may damage the fruit. Phytophthora: the disease can be prevalent during long-term storage at low temperature if orchard control systems are not in operation. Phytophthora is characterised by circular translucent lesions, which develop with grey surface mycelium, and is particularly apparent around the peduncle.

Post-harvest control is achieved by treatment with 0.1% Previncur N. Low Temperature: Storage of unripe papaya below temperatures of 10°C will result in chilling injury. The symptoms are indicated by surface pitting, discolouration of the peel and the flesh, incomplete ripening, poor flavour and increased susceptibility to disease incidence. Ripe papayas will store successfully at lower temperatures, but transport of ripe fruit (more than 50% yellow colour) is not recommended due to the susceptibility to mechanical damage and bruising.

Handling in Importing Countries

For the maximum marketing period of papaya, the fruit should be stored at 10° to 12°C. Temperatures below this range will cause chilling injury and rapid deterioration in fruit quality. To develop ripening in papaya, fruits should be stored at 18° to 25°C and treated with ethylene gas at 100 ppm (0.01%) for 24 hours. Unripe papaya are sensitive to ethylene and will commence ripening if stored with ethylene-producing commodities. Similarly, ripe papaya produce ethylene and will cause deterioration in ethylene-sensitive crops.

5

Practices on Soil Fertility

An experiment was carried out with sugar cane at the Goat and Rabbit Research Centre, Hatay, Vietnam, from January 1993 to December 1995 to study the effect of row width (75, 100 or 150 cm), plant material (stem cuttings or tops), and removing or leaving the dead leaves, on biomass yield and juice quality (°Brix).

Yield of cane stalks was higher when stem cuttings, rather than tops, were used as planting material. The effect was most pronounced in the first harvest year and became less with succeeding harvests. The cost of seed was also reduced by 9-10% compared with the traditional way used by the farmers in North Vietnam. Increasing population density by reducing inter-row spacing to 75 cm led to higher yields of biomass with no reduction in?Brix or extraction rate of the juice. Mulching with the dead leaves increased yields, the difference being more marked in the second than in the first year and even more apparent in the third year (6.3, 20 and 30 % increases for mulching in plant, 1st and 2nd ratoons, respectively). Mulching also improved soil fertility and increased the amount of carbon sequestered in the soil, indicating that mulching is a positive alternative to the traditional way of removing the dead leaves. Soil fertility increased steadily with increasing ratoons indicating that the growing of sugar cane does not exploit soil nutrients but, in contrast, has beneficial effects on growth of subsequent crops.

In recent years a considerable amount of research has been directed at increasing biomass quantity and quality of agricultural crops used as animal feed. Sugar cane is a perennial crop which can be used for animal feed as well as sugar production (Preston and Murguetio 1992). It has a high leaf area index and a high photosynthetic efficiency under strong sunshine, more than any other crop in the tropics (Bassham 1978). The individual and combined effects of certain management practices— planting date, row spacing, planting depth, fertilizer rate, pest control and irrigation — have a great impact on the growth and yield of sugar cane. The number of plants at harvest time is a major determinant of biomass yield, and a density of 50-70,000 plants/ ha was recommended by Tran Van Soi (1988). This was a general recommendation to account for factors such as variety, branching ability, climate and soil conditions. If the plants are too close, there may be too many shoots which will reduce the efficiency of the parent plants and class 1 branches while too large a space between rows will lead to a waste of the area and of solar energy (Duong Duc Thang 1991). In the hilly land of Vietnam, higher production was obtained with the same plant density but arranged in narrower rows (Nguyen Huy Uoc 1987) compared to the traditional method of planting. Other researchers also reported higher yields from narrow spacing (90 and 60 cm) compared with wide spacing (130, 140 and 180 cm) (Sharma 1982; Irvine *et al* 1984; Singh and Singh 1984; Gonzalaz *et al* 1989; Arvind Misra *et al* 1990).

Soil organic matter is an important feature of soil fertility. Ways of increasing soil organic matter are by growing a cover crop (green manure) or by mulching with compost or crop residues. These practices also help to control weed growth. Mulching has been shown to increase the yield of tomatoes and tuber plants (Kaniszewski 1994; Pakyurek *et al* 1994, Khalak and Kumaraswamy 1993; Shin *et al* 1993). The method gave higher yields in fruit trees, vegetables and crops such as mango, sweet pepper, bananas, guava and maize (Farre *et al* 1993; Siwek *et al* 1994; Singh and Singh 1992; Sarad-Gurung *et al* 1994; Sandhu *et al* 1992). Mulching by crop residues was superior to polyethylene sheets (used to control weeds) in terms of incremental cost-benefit ratios (Khalak and Kumaraswamy 1993). There are several reports of improvement of soil fertility as a result of mulching with crop residues (Sandhu *et al* 1992; Kitou and Yoshida 1994; Arzeno 1992).

The hypotheses to be tested in the present study were:

- Reducing row distance would lead to increased yield of biomass
- Planting with stem cuttings rather than the growing point would increase the longevity and the yield of sugar cane
- Returning the dead leaves to the soil (mulching) would increase biomass yield and soil fertility

Materials and Methods

Location: The experiment was carried out from January 1993 to December 1995 at the Goat and Rabbit Research Centre in Bavi district in a hilly area some 60 km North-West of Hanoi with an average slope of 3-8°. The soil composition at the experimental site was as follows: pH 5.3, K_2O 0.06%, P_20_5 0.09%, N 0.14%. In general, the soil is of low fertility and the organic matter has been reduced to very low levels by erosion. Drought occurs frequently in the area.

Treatments and Design: The treatments in a split-plot design with three replications were:

- Planting distances of 75, 100 or 150 cm between rows
- Planting material: Tops (growing points of sugar cane) or stem cuttings (the stalk was cut into pieces of 50 cm)
- Mulching (returning the dead leaves to the soil surface) or no mulching (taking away the dead leaves)
- Harvesting at 10 months or 12 months after planting

Row spacing was the main plot and plant material and mulching were the sub-plots. The area for each experimental plot was 90, 80 and 75 m^2 for row spacings of 150, 100 and 75 cm, respectively. The size of the recorded area was 60 m^2 for all treatments. The total number of plots was 36, giving a total experimental area of 3,569 m^2 (2,940 m^2 for recording data). The plant material was set in double continuous rows with 8, 10 and 12 tonnes/ha for 150, 100 and 75 cm spacing between rows, respectively.

Cultural Practices: Fertilizer was applied according to the normal practice by farmers in the area. The amounts of cattle manure and chemical fertilizers applied.

Management of the Dead Leaves

The dead leaves were collected and weighed every month beginning 6 months after planting. For the mulching treatment the leaves were returned to the soil after they were weighed.

Harvesting

At harvest the experimental plots were divided into 2 parts for harvesting. One half was harvested at 10 months after planting and the other half at 12 months. The whole plant was cut at ground level. The stalk was separated from the whole plant by cutting immediately below the second node measured from the top. The growing points (tops) were then separated from the leaf blades (green leaf). Each component was weighed and sampled for chemical analysis.

Extraction of Juice

Samples of the stalk from each plot (about 10 kg) were crushed by passing them three times through a 2-roll mill driven by a buffalo. On the second and third pass the partially pressed stalks were doubled to maximise extraction of the juice. Extraction rate was expressed as weight of juice as a percentage of the weight of cane stalks. The total soluble solids in the juice (°Brix) were determined using a hand refractometer.

Biological Test of Soil Fertility

Soil samples (from 0-20 cm depth) were taken from each experimental plot immediately after each harvest. Equal amounts (3 kg) were put into clay pots (about 5 litre capacity) for a biological test of overall soil fertility. Three seeds of maize were planted. After 5 weeks the maize plants were removed from the soil, washed to remove soil from the roots and allowed to dry for 1 hour. The total fresh biomass and the maize roots were weighed (Maria Elena Gomez 1993 personal communication).

Soil Analysis

Samples of soil (from 20 cm depth) were taken from each plot after the 12 month harvest. The 18 samples corresponding to the mulching treatments were bulked and 6 sub-samples taken for analysis. The samples from the 18 non-mulching treatments were treated in the same way. The sub-samples were analysed for pH, N, P, K and carbon by standard methods (AOAC 1985). Estimates were made of the populations of fungi, bacteria and actinomycetes according to "Standard methods of analysis for Soil, Plant tissue, Water and Fertilizer" (Philippine-Los Banos, Laguna 1980) with asparagine-manitolagar medium for bacteria, glycerol-agar for actinomycetes and peptone-dextrose-agar plus rose bengal and streptomycin for fungi.

Lemon

Citrus Limon

The leading acid citrus fruit, because of its very appealing colour, odour and flavour, the lemon, *Citrus limon* Burm. f. (syns. C. *limonium* Risso, C. *limonia* Osbeck, *C. medica* var. *limonium* Brandis), is known in Italy as *limone;* in most Spanish-speaking areas as *limon, limon agria, limon real,* or *limon frances;* in German as *limonen;* in French as *citronnier;* in Dutch as *citroen.* In Haiti, it is *limon France;* in Puerto Rico, *limon amarillo.* In the Netherlands Antilles, *lamoentsji,* or *lamunchi,* are locally applied to the lime, not to the lemon as strangers suppose. The lemon is not grown there.

Several lemon-like fruits are domestically or commercially regarded as lemons wherever they are grown and, accordingly, must be discussed under this heading. These include: Rough lemon (C. *jambhiri* Lush.), Sweet lemon *(C. limetta* Risso), 'Meyer' (lemon X mandarin hybrid); 'Perrine' (lime X lemon hybrid); 'Ponderosa' (presumed lemon X citron hybrid), qq.v. under "Varieties".

The true lemon tree reaches 10 to 20 ft. (3-6 m) in height and usually has sharp thorns on the twigs. The alternate leaves, reddish when young, become dark-green above, light-green below; are oblcng, elliptic or long-ovate, 2 1/2 to 4 1/2 in (6.25-11.25 cm) long, finely toothed, with slender wings on the petioles. The mildly fragrant flowers may be solitary or there may be 2 or more clustered in the leaf axils. Buds are reddish; the opened flowers have 4 or 5 petals 3/4 in (2 cm) long, white on the upper surface (inside), purplish beneath (outside), and

20-40 more or less united stamens with yellow anthers. The fruit is oval with a nipple-like protuberance at the apex; 2 3/4 to 4 3/4 in (7-12 cm) long; the peel is usually light-yellow though some lemons are variegated with longitudinal stripes of green and yellow or white; it is aromatic, dotted with oil glands; 1/4 to 3/8 in (6-10 mm) thick; pulp is pale-yellow, in 8 to 10 segments, juicy, acid. Some fruits are seedless, most have a few seeds, elliptic or ovate, pointed, smooth, 3/8 in (9.5 mm) long, white inside.

Origin and Distribution: The true home of the lemon is unknown, though some have linked it to northwestern India. It is supposed to have been introduced into southern Italy in 200 A.D. and to have been cultivated in Iraq and Egypt by 700 A.D. It reached Sicily before 1000 and China between 760 and 1297 A.D. Arabs distributed it widely in the Mediterranean region between 1000 and 1150 A.D. It was prized for its medicinal virtues in the palace of the Sultan of Egypt and Syria in the period 1174-1193 A.D. Christopher Columbus carried lemon seeds to Hispaniola in 1493. The Spaniards may have included lemons among the fruits they introduced to St. Augustine. They were grown in California in the years 1751-1768. Lemons were reported to be increasingly planted in northeastern Florida in 1839. Because of heavy imports from Sicily, commercial culture in Florida and California was begun soon after 1870 and grew to the point where 140,000 boxes were being shipped out of Florida alone. The small Florida industry was set back by a freeze in 1886, the susceptibility of the lemon to scab, and the unfavorable climate for curing the fruit, and also competition from California. Following the devastating freeze of 1894-95, commercial lemon culture was abandoned in Florida. Not until 1953 was interest in lemon-growing revived in Central Florida to take advantage of the demand for frozen concentrate and for natural cold-press lemon oil. At that time, Florida was importing lemons from Italy for processing. Plantings grew to 8,700 acres by 1975. Freezes caused 50% reduction by 1980. Still, in 1984, Florida exported $2 million worth of lemons.

In the meantime, Arizona had developed lemon orchards, though on a smaller scale than California. In the 1956-57 season, California produced 11 million gallons (42 million liters) of frozen lemon concentrate while Florida's output was still very small. California and Arizona became the leading sources of lemons in the western hemisphere. In recent years, California has produced nearly double the crop that can be profitably marketed fresh or processed. Foreign competition has increased and many California growers have destroyed their lemon groves or topworked the trees to oranges, but new cultural techniques making summer production possible may reverse the trend.

Guatemala has in the past 2 decades developed commercial lemon culture, primarily to produce the peel oil for its essential oil industry and secondarily for the purpose of dehydrating the fruit and preparing a powder for reconstituting into juice. Southern Mexico, too, is now a major grower of lemons, also primarily for lemon peel oil. Lemons are rarely grown for the fresh fruit market in Latin America. In South America, Argentina leads in lemon culture with Chile a distant second. Among the world's leading lemon growers and exporters are Italy, Spain, Greece, Turkey, Cyprus, Lebanon, South Africa and Australia. Lemons can be grown only at medium and high elevations in the Philippines.

Varieties: With the resumption of lemon-growing in Florida, workers at the Citrus Experiment Station, Lake Alfred, began a search for the most suitable cultivars, whether in dooryards, or in the United States Department of Agriculture planting at Orlovista, or the Lake Alfred collection. By late 1950, 200 selections had been brought together from various parts of the United States. Of these, 40 were budded onto 30-year-old grapefruit trees on rough lemon rootstock on the Minute Maid property at Avon Park. Two selections grown elsewhere were included in the studies-evaluation for thorniness, cold-and disease-susceptibility, sizes, juiciness, flavour, number of segments and seeds, yields, and quality of peel oil. The majority of the selections were judged undesirable; only a few showed promise for processing and fresh fruit marketing purposes. For processing, 'Villafranca' rated highest, followed by 'Eustis', 'Bearss', 'Perkin' and 'Avon'. Any of these, properly harvested and cured would be suitable for marketing fresh. Libby, McNeil & Libby, when planning for their lemon orchard at Babson Park, Florida, about 1948, tested varieties from all major lemon-producing areas of the world and chose 'Bearss' as rating highest in quality and quantity of juice, which was their chief concern at the time. In 1960, they added marketing of the fresh fruit and found the 'Bearss' equally desirable for this purpose.

The following are brief descriptions of most of the better known cultivars of true lemons and of lemon-like fruits that are accepted as lemons in home or commercial usage, and a few of the lesser-known.

'Armstrong' ('Armstrong Seedless')–a sport discovered in a private grove at Riverside, California, about 1909. Patented in 1936 by Armstrong Nurseries. Resembles 'Eureka' except that it usually bears seedless or near-seedless fruits. If planted among other lemon trees will occasionally have a few seeds.

'Avon'–first noticed as a budded tree in Arcadia, Florida. A budded tree propagated from the original specimen around 1934 was planted in the Alpine Grove in Avon Park; it produced heavy crops of fruits highly suitable for frozen concentrate. It, therefore, became the source of budwood for commercial propagation by Ward's Nursery beginning in 1940.

'Bearss' ('Sicily', but not the original introduction by Gen. Sanford in 1875, which has disappeared)–a seedling believed to have been planted in 1892, discovered in the Bearss grove near Lutz, Florida, about 1952. Closely resembles 'Lisbon'. It is highly susceptible to scab and greasy spot and oil spotting. The tree is vigorous and tends to produce too many water sprouts. Nevertheless, it has been propagated commercially by Libby, McNeill & Libby since 1953 because the peel is rich in oil. It constitutes 20% of Brazil's lemon/lime crop.,

'Berna' ('Bernia', 'Vema', 'Vernia')–oval to broad-elliptic, with pronounced nipple, short neck; peel somewhat rough, medium-thick, becoming thinner in summer, tightly clinging. Seeds generally few or absent. Ripens mostly in winter; fruits keep well on tree until summer but become too large. Tree is vigorous, large, prolific. This is the leading cultivar of Spain and important in Algeria and Morocco. It is too much like the

'Lisbon' to be of value in California. In Florida, it has been found deficient in acid, low in juice, and too subject to scab.

'Eureka'–originated from seed taken from an Italian lemon (probably the 'Lunario') and planted in Los Angeles in 1858; selected in 1877 and budwood propagated by Thomas Garey who named it 'Garey's Eureka'. The fruit is elliptic to oblong or rarely obovate, with moderately protruding nipple at apex, a low collar at the base; peel yellow, longitudinally ridged, slightly rough because of sunken oil glands, medium-thick, tightly clinging; pulp greenish-yellow, in about 10 segments, fine-grained, tender, juicy, very acid. Fruits often borne in large terminal clusters unprotected by the foliage. Bears all year but mostly late winter, spring and early summer when the demand for lemons is high. Tree of medium size, almost thornless, early-bearing, prolific; not especially vigorous, cold-sensitive, not insect-resistant; relatively short-lived. Not suitable for Florida. Grown commercially in Israel. One of the 2 leading cultivars of California, though now being superseded by clonal selections with more vigor, *e.g.*, 'Allen', 'Cascade', 'Cook', and 'Ross'. 'Lambert Eureka' is a chance seedling found in 1940 on the property of Horace Lambert in New South Wales. It is vigorous and productive.

'Femminello Ovale'–one of the oldest Italian varieties; short-elliptic with low, blunt nipple; slightly necked or rounded at base; of medium size; peel yellow, finely pitted, medium-smooth, medium-thick, tightly clinging; pulp in about 10 segments, tender, juicy, very acid, of excellent quality, with few, mostly undeveloped, seeds. Fruits all year but mainly in late winter and spring; ships and stores well. The tree is almost thornless, medium-to very-vigorous, but highly susceptible to *mal secco* disease. This is the leading cultivar in Italy, accounting for 3/4 of the total lemon production, and 1/5 of the crop is processed as single-strength juice.

'Genoa'–introduced into California from Genoa, Italy, in 1875. Almost identical to 'Eureka'; ovoid or ovate-oblong with blunt nipple at apex; base rounded or slightly narrowed; of medium size; peel yellow, medium-thick, tightly clinging; pulp in 10-12 segments, melting, medium-juicy, with 29 to 51 seeds which are light-brown within. Tree is shrubby, nearly trunkless, spreading, very thorny, cold-hardy. Grown commercially in India, Chile and Argentina.

'Harvey'–of unknown parentage; was found by Harvey Smith on the property of George James in Clearwater, Florida. Fruit much like 'Eureka'. Tree highly cold-tolerant, compatible with several rootstocks. Commercially propagated by Glen St. Mary Nurseries Company, near Jacksonville, Florida, since 1943.

'Interdonato' ('Special')–a lemon X citron hybrid that originated on property of a Colonel Interdonato, Sicily, around 1875; oblong, cylindrical, with conical, pointed nipple at apex, short neck or collar at base; large; peel yellow, smooth, glossy, thin, tightly clinging; pulp greenish-yellow, in 8 or 9 segments, crisp, juicy, very acid, faintly bitter. Very few seeds. Earliest in season; mostly fall and early winter. Tree vigorous, usually thornless, medium-resistant to *mal secco*; of medium yield; accounts for 5% of Italy's crop.

'Lisbon' (perhaps the same as 'Portugal' in Morocco and Algeria)–originated in Portugal, possibly as a selection of 'Gallego'; reached Australia in 1824; first catalogued in Massachusetts in 1843; introduced into California about 1849 and catalogued there in 1853; introduced into California from Australia in 1874 and again in 1875. Fruit almost identical to 'Eureka'; elliptical to oblong, prominently nippled at apex, base faintly necked; peel yellow, barely rough, faintly pitted, sometimes slightly ribbed, medium-thick, tightly clinging; pulp pale greenish-yellow, in about 10 segments, fine-grained, tender, juicy, very acid, with few or no seeds. Main crop in February, second crop in May. Fruit is borne inside the canopy, sheltered from extremes of heat and cold. Tree large, vigorous, thorny, prolific, resistant to cold, heat, wind. Not well adapted to Florida. It is low-yielding and short-lived in India. Surpasses 'Eureka' in California. Has given rise to a number of clonal selections, particularly 'Frost', originated by H. B. Frost at the Citrus Research Station, Riverside, California in 1917 and released about 1950; also 'Prior Lisbon' and the more vigorous 'Monroe Lisbon'.

'Meyer'–a hybrid, possibly lemon X mandarin orange; introduced into the United States as S.P.I. #23028, by the agricultural explorer, Frank N. Meyer, who found it growing as an ornamental pot-plant near Peking, China, in 1908; obovate, elliptical or oblong, round at the base, occasionally faintly necked and furrowed or lobed; apex rounded or with short nipple; of medium size, 2 1/4 to 3 in (5.7-7.5 cm) wide and 2 1/2 to 3 1/2 in (6.25-9 cm) high; peel light-orange with numerous small oil glands, 1/8 to 1/4 in (3-6 mm) thick; pulp pale orange-yellow, usually in 10 segments with tender walls, melting, juicy, moderately acid with medium lemon flavour; seeds small, 8 to 12. Tends to be everbearing but fruits mostly from December to April. Tree small, with few thorns, prolific, cold-resistant; produces few water sprouts, and is only moderately subject to greasy spot and oil spotting. It is easily and commonly grown from cuttings. Does well on sweet orange and rough lemon rootstocks; is not grafted onto sour orange because it is a carrier of a virulent strain of tristeza. Grown for home use in California; in Florida, both for home use and to some extent commercially for concentrate though the product must be enhanced by the addition of peel oil from true lemons, since that from 'Meyer' peel is deficient in flavoring properties. Has been fairly extensively planted in Texas and in Queensland, Australia, and New Zealand.

'Monachello' (Moscatello')–suspected of being a lemon X citron hybrid; elliptical, with small nipple and no neck, merely tapered at apex and base; medium-small; peel yellow, smooth except for large, sunken oil glands, thin, clinging very tightly; pulp in 10 segments, tender, not very juicy, not sharply acid. Bears all year but mainly winter and spring. Tree not vigorous, slow-growing, almost thornless, with abundant, large leaves; bears medium-well, resistant to *mal secco*, and has been extensively planted in Italy in areas where the disease is common.

'Nepali Oblong' (Assam', 'Pat Nebu')–originated in Assam; fruit resembles citron in some aspects; long-elliptic to oblong-obovate, with wide, short nipple; medium-large; peel greenish-yellow, smooth, glossy, medium-thick; pulp greenish-yellow in 11 segments, fine-grained, very juicy, of medium acidity, with few or no seeds. Everbearing. Tree large, vigorous, spreading, medium-thorny, prolific; foliage resembles that of the citron. Commercial in India.

'Nepali Round'–of Indian origin; round, without distinct nipple; juicy; seedless. Tree large, vigorous, compact, nearly thornless, medium-prolific. Successfully cultivated in South India.

'Perrine'–a Mexican lime X 'Genoa'-lemon hybrid created by Dr. Walter Swingle and colleagues in 1909, but still a fairly typical lemon; it is lemon-shaped, with small nipple at apex, necked at base; of medium size; peel pale lemon-yellow, smooth, slightly ridged, thin, tough; pulp pale greenish-yellow, in 10 to 12 segments having thin walls; tender, very juicy, with slightly lime-like flavour but acidity more like lemon; seeds usually 4 to 6, occasionally as many as 12, long-pointed. Everbearing. Tree cold-sensitive but less so than the lime; resistant to wither tip and scab but prone to gummosis and other bark diseases. In the early 1930's, was extensively planted in southern Florida on rough lemon rootstock, but no longer grown.

'Ponderosa' ('Wonder'; 'American Wonder')–a chance seedling, possibly of lemon/citron parentage, grown by George Bowman, Hagerstown, Maryland around 1886 or 1887; appeared in nursery catalogs in 1900 and 1902; obovate, lumpy and faintly ribbed, slightly necked at base; large, 3 1/2 to 4 1/8 in (9-11 cm) wide, 3 1/2 to 4 3/4 in (9-12 cm) high; peel light orange-yellow, with medium-large oil glands, flush or slightly depressed; 3/8 to 1/2 in (1-1.25 cm) thick; pulp pale-green, in 10 to 13 segments with thick walls; juicy, acid; seeds of medium size, 30 to 40 or more, brown within. Everbearing. Tree small, moderately thorny; buds and flowers white or barely tinged with red-purple. More sensitive to cold than true lemons. Grown for home use and as a curiosity in California and Florida and in small-scale commercial plantings since 1948. Rather widely cultivated as an indoor potted plant in temperate regions.

'Rosenberger'–a clone found in a grove of 'Lisbon' and 'Villafranca' trees at Upland, California; was planted in the Rosenberger orchard and gained recognition as a superior cultivar. Tree closely resembles that of 'Villafranca'. Fruit is somewhat like 'Lisbon' but is shorter and broader and less tapered at base. Tree vigorous and prolific. Became popular in California in the 1960's.

'Rough Lemon' ('Florida Rough'; French'; 'Mazoe'; Jamberi')–perhaps a lemon X citron hybrid, but has been given the botanical name of C. *jambhiri* Lush. Believed to have originated in northern India, where it grows wild; carried in 1498 or later by Portuguese explorers to southeastern Africa where it became naturalized along the Mazoe River; soon taken to Europe, and brought by Spaniards to the New World; is naturalized in the West Indies and Florida; oblate, rounded or oval, base flat to distinctly necked, apex rounded with a more or less sunken nipple; of medium size, averaging 2 3/4 in,(7 cm) wide, 2 1/2 (6.25 cm) high; peel lemon-yellow to orange-yellow, rough and irregular, with large oil glands, often ribbed; 3/16 to 3/8 in (5-10 mm) thick; pulp lemon-yellow, usually in 10 segments, medium-juicy, medium-acid, with moderate lemon odour and flavour; seeds small, 10 to 15, brownish within. Reproduces true from seeds, which are 96% to 100% nucellar. Tree large, very thorny; new growth slightly tinged with red; buds and flowers with red-purple. The scant pulp and juice limit the rough lemon to home use. It is appreciated as a dooryard fruit tree in Hawaii and in other tropical and subtropical areas where better lemons are not available. The tree has been of great importance as a rootstock for the sweet orange, mandarin orange and grapefruit. It is not now used as a rootstock for lemon in Florida because of its susceptibility to "blight" (young tree decline). It is also prone to Alternaria leaf spot (*Alternaria citri*) in the nursery, to foot rot (*Phytophthora parasitica*). Incidence varies with the clone and certain clones show significant resistance. In trials at Lake Alfred, 3 atypical clones showed immunity to leaf spot, while a typical rough lemon clone, 'Nelspruit 15', from South African seed, proved highly resistant to leaf spot and also extremely cold tolerant.

'Santa Teresa'–an old tree discovered to be disease-free in a 'Fermminello Ovale' orchard in Italy that had been devastated by *mal secco*. Budded trees from the original specimen were being commonly planted in the 1960's wherever the disease was prevalent in Italy.

Sweet Lemon (C. *limetta Risso)–a* general name for certain non-acid lemons or limettas, favoured in the Mediterranean region, In India, they are grown in the Nilgiris, Malabar and other areas. The fruits are usually insipid, occasionally subacid or acid. The seeds are white within and the tree is large, resembling that of the orange. One cultivar, called 'Dorshapo' after the plant explorers, Dorsett, Shamel and Popenoe, who introduced it from Brazil in 1914, resembles the 'Eureka' in most respects except for the lack of acidity. Another, called 'Millsweet', apparently was introduced into California from Mexico and planted in a mission garden. It was reproduced at the old University of California Experiment Station at Pomona. Neither is of any commercial value.

'Villafranca'–believed to have originated in Sicily; introduced into Sanford, Florida, from Europe around 1875 and later into California. Closely resembles 'Eureka'; of medium size. Tree is more vigorous, larger, more densely foliaged, and more thorny than 'Eureka' but becomes thornless with age. One strain is everbearing; another fruits heavily in summer. This was the leading lemon cultivar in Florida for many years; is cultivated commercially in Israel; is low-yielding and short-lived in India. It is little grown in California but has given rise to certain selections that are of importance, particularly 'Galligan Lisbon' and 'Corona Foothill Eureka'.

Climate: Because of its more or less continuous state of growth, the lemon is more sensitive to cold than the orange and less able to recover from cold injury. The tree is defoliated at 22° to 24° F (-5.56°-4.44° C). A temperature drop to 20° F (-6.67° C) will severely damage the wood unless there has been a fortnight of near-freezing weather to slow down growth. Flowers and young fruits are killed by 29° F (-1.67° C) and nearly mature fruits are badly damaged below 28° F (-2.22° C). On the other hand, the lemon attains best quality in coastal areas with summers too cool for proper ripening of oranges and grapefruit. Therefore, the lemon has a relatively limited climatic range. In Florida, lemons are produced commercially as far north as Ft. Pierce on the East Coast and Ruskin on the West Coast. The 'Meyer' lemon, as a dooryard tree, can be grown wherever oranges thrive, even as far west as Pensacola.

The fruits are scarred and the tree readily defoliated by winds, and benefit by the protection of windbreaks. Lemons are grown in both dry and humid atmospheres, the latter being a disadvantage mainly in the processes of curing and storing. Over a large lemon-growing region in California, annual rainfall varies from 25 to 125 cm. In long, dry periods, the lemon must be irrigated.

Soil: The lemon tree has the reputation of tolerating very infertile, very poor soil. In Florida, groves are mostly on sand. In California, excellent growth is maintained on silty clay loam of high water-holding capacity. In Guatemala, recommended soils are sand, clay and sandy-clay-deep, with high permeability and good drainage. Black soils are also suitable if not lying over calcareous subsoil. Ph should be between 5.5 and 6.5. If acidity is high, it is necessary to apply lime to achieve the optimum level.

Propagation: The rough lemon is widely grown from seed. The 'Meyer' lemon is easily reproduced by rooting large cuttings in the nursery and planting them directly in the grove. They fruit 2 to 3 years sooner than budded trees and have a long life, remaining in full production for over 30 years, perhaps much longer.

In Florida, commercial lemons have been budded onto 'rough lemon', sweet orange, and 'Cleopatra' mandarin rootstocks. More recent practices are the utilization of sour orange, Volkamer lemon (C. *volkameriana),* and alemow (C. *macrophylla* Wester, an old Philippine lemon/pummelo hybrid). The latter is employed in California on soils containing an excess of soluble salts and boron. If citranges are used as rootstocks for 'Eureka', bud union crease will kill the tree.

Culture: Lemon trees should be spaced 25 ft. (7.6 in) apart each way. If crowded or "hedged", production declines. The trees must be pruned when young and kept below 10 or 12 ft. (3-3.6 m) in height. They are cut back severely after 12 years or replaced. Weeds must be controlled but lemon trees are very sensitive to herbicides.

In Florida, fertilizing may be done 3 times a year between mid-November and the end of April, at the gradually increasing rate of 4 to 10 lbs (1.8-4.5 kg) per tree up to an age of 50 years. Nitrogen and potash are given in equal amounts under normal soil conditions. A nutritional spray with copper added is applied after spring bloom. Fertilizer and irrigation programmes should be varied according to the desired goal: fresh fruit marketing or processing. High nitrogen steps up yield and peel oil content but also results in more scab infection and poor curing. Potash increases acidity. Heavy irrigation increases yield and peel oil, scab infection, size of fruit and accelerates maturity.

In California, foliar spraying of urea is preferred over ground application of nitrogen which can lead to accumulations of salts and also contamination of groundwater. Leaf analyses are made to determine the nitrogen requirements of each cultivar for maximum yield. 'Eureka', in a 6-year test, showed no response to increased levels of nitrogen. In New Zealand, mature trees (15 to 20 years old) are given 25 to 30 lbs (11.3-13.6 kg) of complete mixed fertilizer annually, also heavy dressings of organic manure or mulch.

In Sicily, growers have, for over 50 years, made a practice of withholding water in summer-for 35 to 60 days-until the trees begin to wilt. Then the trees are heavily irrigated and given high nitrogen fertilizer which induces a second bloom in August or early September, producing a crop the following summer when lemons are scarce and prices are high. This system, called the "Verdelli process", was adopted on a little over 1,000 acres (405 ha) in California in 1983. Adequate bloom did not occur on sandy or shallow soils, but 80% of the plantings on gradually dehydrating, fine-textured soil bloomed well. Nearly $3 million was expected from this extra crop of summer lemons in the Central Valley and the Riverside area in 1984. New horticultural techniques are needed to overcome the handicaps of higher use of fertilizer, increased insect and fungus problems, effects of moisture stress on fruit quality, and low temperature hazard to immature fruits in winter.

In 1965, a team of California horticulturists initiated experimental trellis culture of 'Prior Lisbon' lemon on C. *macrophylla* rootstock. It was found that the labour of training, and repeated pruning either manually or by machine hedging and topping, was excessive and uneconomic. Guatemalan and Mexican growers interplant short-term crops such as beans, cassava, yautia *(Xanthosoma),* in the rainy season, and tomatoes and peppers during the winter when the lemon trees will be irrigated and fertilized.

Harvesting and Handling: The marketability of lemons depends on the stage at which they are picked. Italian lemons for export are harvested as early as possible and are naturally "cured" in transit. In early days, California and Florida lemons were allowed to remain on the trees until they became too large. It was realized that early picking is necessary and California and Arizona adopted the practices of picking at any time after the fruits reach a 25 % juice content, and using rings to gauge the commercially acceptable size, and repeated spot-picking with clippers. Mechanical picking is impossible with lemons. The fruits are highly prone to oil spotting (oleocellosis) and cannot be handled roughly nor picked wet.

Formerly, Florida lemons were picked from mid-July to October for shipping fresh, and the balance in November was harvested for processing. Lemons under 2 1/8 in (5.4 cm) are too immature to attain proper quality for marketing and fruits over 2 1/2 in (6.25 cm) are too large. Manual spot-picking has been commonly practiced, but some producers have found it too costly, and are harvesting the entire crop at one time and grading for fresh sale or processing in the packing-house, discarding all undersized fruits. The lemons, after sorting according to colour, washing and coating with a fungicide and a thin layer of wax are stored (cured) until ready for shipping.

Yield: Lemon tree yields vary considerably with the cultivar, the location and weather conditions. A yield of 3 boxes per tree is commercially satisfactory in Florida. In India, a 6-year-old tree bore 966 fruits and, at 9 years of age, had produced a total of 3,173 fruits.

Storage: Florida's climate is unfavourable for long-term curing. It has been claimed that a 10-day curing period is adequate and degreening of Florida fruit is not needed. A major producer keeps the newly harvested fruits for 48 hours at 60° F (15.56° C) and 95% humidity, then passes them through a pre-grading procedure to eliminate all that are unusable. The usable fruits are then treated with fungicide against stem-end rot and returned to the curing room. Those harvested early in the season need 3 weeks to colour-up, the last may require less than a week. Finally, the fruits are washed, given a second fungicidal treatment, dried, waxed and packed.

Generally, lemons are cured at 56° to 58° F (13.33°-14.4° C) and 85-90% relative humidity. Green fruits may be held for 4 months or more, while the peel becomes yellow and thinner, the pulp juicier (6-80%) and the proportion of soluble solids higher (7-24%). Sometimes the degreening process is hastened by exposing the fruit to ethylene gas, ethephon, or silane, but this practice tends to stimulate decay, mainly through the shedding of the "button" (stem stub), the absence of which allows entry of *Diplodia natalensis, Phomopsis citri,* or *Alternaria* mycelium. Various auxins have been studied to determine which can be applied before storage to prevent button loss without delaying degreening. In 1982, Israeli investigators reported that decay losses from degreening procedures can be greatly reduced (from over 50% to 6.3%) by packaging the fruits in 10 micrometer-thick high-density polyethylene. This treatment makes it possible to store lemons with minimum damage for as long as 6 months.

In the past, New Zealand lemons for storage have been individually wrapped in diphenyl-treated paper after washing and dipping in a 200 ppm solution of 2,4,5-T and then waxing. The fruits were marketable after storing for 4 months at room temperature. Lemons can be kept for weeks in the home refrigerator if placed in a jar with a tight-fitting lid to prevent loss of moisture.

Lemons for export from Florida to Hawaii and Arizona must be fumigated with methyl bromide because of possible infestation by the Caribbean fruit fly. For sale within the state, other methods must be employed.

Pests and Diseases: In Southeast Asia, many species of ants attack the root system and the farmer times the opening of the water gates so as to force the ants to the surface of the beds, where he burns them with fire.

One of the 3 most serious arthropod pests of the lemon and other citrus trees in California is California red scale, *Aonidiella aurantii.* In the southern part of the state it is under biological control but it requires applications of pesticides in the San Joaquin Valley. In Florida, rust mites, purple mites and purple scale may at times be troublesome but they are all controllable with appropriate sprays.

Young lemon trees in California sometimes require protection from wild rabbits.

Diseases are the greater challenges. In Florida, the main lemon diseases are scab *(Elsinoe fawcetti)* on fruit, leaves and twigs; anthracnose of fruit (stylar-end-rot), leaves and twigs caused by both *Colletotrichum gloeosporioides* and *Glomerella cingulata;* greasy spot *(Mycosphaerella citri* or *Cercospora citri-grisea);* and gummosis *(Diaporthe citri).* The latter organism also causes melanose and die-back, and stem-end rot. Stem-end rot may also arise from attack by *Botryosphaeria ribis* and *Diplodia natalensis.*

Other lemon diseases recorded in Florida are branch knot *(Sphaeropsis tumefaciens),* damping-off *(Rhizoctonia solani),* leaf spot *(Mycosphaerella horii, Alternaria citri,* and *Catenularia sp.;* algal leaf spot or green scurf *(Cephaleuros virescens);* tar spot *(Cercospora gigantea);* felt fungus *(Septobasidium pseudopedicellatum);* charcoal root rot *(Macrophomia phaseolina);* root rot *(Fusarium oxysporum, Pythium ultimum,* and *Phytophthora parasitica;* heart rot and wood rot *(Fomes applanatus* and *Ganoderma sessilis);* crinkly leaf and exocortis viruses; and green mold *(Penicillium digitatum);* blue mold (P. *italicum);* and pink mold (P. *roseum).* In 1955, the lemon budwood certification programme was begun to provide virus-free stock for growers.

Red algae infests lemon trees and causes much dieback unless controlled with copper fungicide in the summer. Zinc deficiency causes stunting of twigs, reduced flowering, premature dropping of fruit, and yellow bands along the leaf veins. Manganese deficiency is evidenced by interveinal chlorosis and subsequent necrosis, shedding of leaves, flowers and young fruit. In India, fruit cracking occurs when dry periods are followed by heavy rains. Cracking can be largely avoided by frequent light irrigation during the dry period and early picking.

Stored lemons are subject to the stem-end rots and the molds listed above. The albedo may show small dark sunken areas even though this defect is not visible externally. Cultivars differ in their ability to resist decay.

Food Uses: Slices of lemon are served as a garnish on fish or meat or with iced or hot tea, to be squeezed for the flavourful juice. In Colombia, lemon soup is made by adding slices of lemon to dry bread roll that has been sauteed in shortening until soft and then sieved. Sugar and a cup of wine are added and the mixture brought to a boil, and then served.

Lemon juice, fresh, canned, concentrated and frozen, or dehydrated and powdered, is primarily used for lemonade, in carbonated beverages, or other drinks. It is also used for making pies and tarts, as a flavoring for cakes, cookies, cake icings, puddings, sherbet, confectionery, preserves and pharmaceutical products. A few drops of lemon juice, added to cream before whipping, gives stability to the whipped cream.

Lemon peel can be candied at home and is preserved in brine and supplied to manufacturers of confectionery and baked goods. It is the source of lemon oil, pectin and citric acid. Lemon oil, often with terpenes and sesquiterpenes removed, is added to frozen or otherwise processed lemon juice to enrich the flavour. It is much employed as a flavoring for hard candies.

Toxicity: The thorns of the lemon tree inflict painful punctures and scratches. Lemon peel oil may cause contact dermatitis, chronic in those who handle, cut and squeeze lemons daily. Parts of the body touched by contaminated hands may show severe reactions after exposure to the sun. People that suck lemons may suffer irritation and eruptions around the mouth. The wood of lemon trees and its saw-dust may induce skin reactions in sensitive woodworkers.

Other Uses

Lemon juice is valued in the home as a stain remover, and a slice of lemon dipped in salt can be used to clean copper-bottomed cooking pots. Lemon juice has been used for bleaching freckles and is incorporated into some facial cleansing creams. Lemon peel oil is much used in furniture polishes, detergents, soaps and shampoos. It is important in perfume blending and especially in colognes.

Petitgrain oil (up to 50% citral), is distilled from the leaves, twigs and immature fruits of the lemon tree in West Africa, North Africa and Italy. With terpenes removed, it is greatly prized in colognes and floral perfumes.

Lemon peel, dehydrated, is marketed as cattlefeed.

Lemonade, when applied to potted plants, has been found to keep their flowers fresh longer than normal. But it cannot be used on chrysanthemums without turning their leaves brown.

Wood: The wood is fine-grained, compact, and easy to work. In Mexico, it is carved into chessmen, toys, small spoons, and other articles.

Medicinal Uses: Lemon juice is widely known as a diuretic, antiscorbutic, astringent, and febrifuge. In Italy, the sweetened juice is given to relieve gingivitis, stomatitis, and inflammation of the tongue. Lemon juice in hot water has been widely advocated as a daily laxative and preventive of the common cold, but daily doses have been found to erode the enamel of the teeth. Prolonged use will reduce the teeth to the level of the gums. Lemon juice and honey, or lemon juice with salt or ginger, is taken when needed as a cold remedy. It was the juice of the Mediterranean sweet lemon, not the lime, that was carried aboard British sailing ships of the 18th Century to prevent scurvy, though the sailors became known as "limeys".

Oil expressed from lemon seeds is employed medicinally. The root decoction is taken as a treatment for fever in Cuba; for gonorrhea in West Africa. An infusion of the bark or of the peel of the fruit is given to relieve colic.

Biological Farming for Sustainable Output

Sustainable modern and biological farming is an elusive and difficult concept to define when one considers the globally rising population and meet its obligations for hunger and health. Generally, biological farming encompasses objectives of maintaining and conserving soil productivity, conserving of conventional and unconventional energy, environmental quality, food quality, resource quality, preserving Bio-diversity and economic viability of suitable agro-systems. Modern farming methods threaten the environment in many ways: soil erosion, water pollution, the destruction of wildlife habitat, deforestation and the deteriorating fertility of the soil itself. These also bring health risks through widespread use of pesticides. As a result, organic methods of cultivation are being revalued, and more widely adopted.

Sustainable production involves a multitude of "need-based" systems ranging from modern tested technology systems for different crops to an organic sustainable production over time and space. Conversions for conventional (inorganic or chemical or modern) sustainable farming systems to sustainable organic farming amounts to low input costs and premiums by virtue of quality and harmless produce. For instance, a relationship exists between farmers and their customers under the cooperation system in Japan who have adopted organic farming. The produce, despite not being so cosmetically attractive and possibly costing more, between three and five million Japanese now regularly buy organic produce for different health reasons.

Within the context of India, although the use of high-yielding varieties has led to an increase in production, it has been at the expanse of increase in chemical use, which, it is argued, is unsustainable. Low production and productivity are the core to problems of Indian agriculture. Despite substantial increases in the use of modern inputs, the yields are much lower than those in advanced countries. Factors impeding the growth of crop yields in India are inadequate irrigation facilities, poor manuring practices, and obsolete methods of cultivation. The components of the integrated nutrients supply system suitable for easy adopting include crop rotation, green manures and bio-fertilisers. Effective water saving, equity in water sharing and efficiency in water delivery and use are indispensable for sustainable management of available surface and ground water resources vis-a-vis sustainable production. To sustainable growth of agriculture and prosperity of farmers in real concerns, the use of self-grown and produced inputs has to be promoted. These inputs must replace the market bought modern inputs.

Agriculture should not damage the environment, livestock rearing, regional programmes for protection of water quality, reversion to organic farming, reduction of livestock density by increasing the fodder area of farms, action for the protection and management of fauna and flora, training to local operations to adapt ecologically sustainable farming systems in vulnerable areas and regulations on reforestation are must. When environmental protection is the aim, the maintenance of the farming population is essential to mitigate the problem of unemployment in a country like India. However, environmentally adapted production systems (such as organic farming) may be transformed into a marketing asset if the products can be processed locally and marketed as a special brand. This cannot usually be done by the individual farmer, but must be done by a group of farmers.

In Australia organic farms have shown the lowest nitrate content in the drain water.

The published information on the production of US maize and potatoes employing different technologies and energy inputs is an indicative that the organic production of grains, such as maize, requires less fossil energy than conventional crop production. For some crops, the organic production system may also produce higher crop yields, as demonstrated with the organic maize system. Combining low inputs with environmentally sound technologies clearly requires that farmers have greater knowledge of the management strategies used in the agro-ecosystem than farmers who employ the more heavy-chemical use systems. For this reason those countries that have active plans to reduce pesticides and commercial fertilisers in agriculture are investing heavily in research and supporting extension education.

Keeping in view the importance of biological farming system, the parameters of research and development — the quantitative and qualitative assessment of costs and returns from alternative farming practices, reducing environmental pollution and health hazards; the inter-dependencies between components of the farming system controlled by the farmer, and interaction of these components with physical, biological and social-economic factors; the inter-disciplinary approach to agriculture research based on an integrated farm production units with limited resources; impetus to investigate and practice organic farming, reduced tillage, methods of conservation of soil as well as integrated pest management, with the overall aim of reducing costs and sustaining yields—must be watched carefully in sustaining any farming system and recommending the best of all.

Super Rice in the Offing

A "Super rice" that can give 25 per cent more yield than all previous paddy breeds is all set to enter the market within the next five years.

Developed by the International Rice Research Institute (IRRI) in the Philippines, the new variety could break the yield plateau reached by modern high-yielding rice varieties, according to Mr Gurdev S. Khush, who is leading a team at the IRRI to develop the variety.

"Field testing is currently going on in China and in the Philippines, "Khush said, adding the super rice was expected to be in the fields by 2003.

Khush was addressing a press conference after signing an agreement between the IRRI and the Indian Council of Agricultural Research (ICAR) for collaboration in rice research.

"The IRRI has developed about 400 different varieties of super rice. They will be tested in India in association with the ICAR within the next two years to find out which one suits the Indian soil and climatic conditions best," he said.

The super rice, developed through cross-hybridisation of two aptly chosen parent plants, had a plant architecture that was completely different from all previous plants, Khush said.

It had fewer but sturdier stems, all of them bearing panicles with panicle housing 200-250 more grains than previous varieties.

It took 100-130 days for growth and maturation, Mr Khush said. The dark green-coloured plant had erect leaves to increase photosynthetic effectiveness.

It had an organised root system that did not allow wastage of water and micro-nutrients needed for plant growth, Khush said.

IRRI scientists were currently working to make it disease resistant by hybridising it with other rice varieties that could resist many of the common plant virus and other insects.

"We are now improving the super rice by building multiple pest and disease resistance into it, "Khush said.

In normal paddy varieties, cereal to biomass ratio was 30:70. But super rice was capable of giving a cereal to biomass ratio of 50: 50 which meant a significant increase in rice productivity.

With ever-increasing population and decrease in land availability and water, rice productivity needs to be increased by 40-50 per cent by the middle of the next century, "he said.

Novel Scheme for Fruit Growers

THE Horticulture Department of Punjab in its endeavour to improve horticulture has come out with a novel scheme. The department proposes to spread its package of practices to the farmers through its horticulture development officers (HDOs). The state has been divided into various "blocks", each headed by an HDO.

Fifteen villages are selected from each block and from these villages 15 farmers are shortlisted who show cooperation and the will to learn and listen.

The department adopts 15 trees of whatever fruit the farmer grows (grapevine in the cases of grapes).

The department has also created "estates" in areas where there is concentration of a particular type of fruit or vegetable. For example, Hoshiarpur is a citrus-growing area and by picking up a central village it covers all the citrus orchards within a 10-km radius. The HDO is required to cover all orchards in the vicinity, contrary to the 15 in blocks. The number of trees adopted varies from estate to estate.

The "thrust", which is also the name of the scheme, is on doing all the activities which are recommended by the department on these 15 plants. The logic behind is to show and train the farmer how proper and timely execution of all farming activities makes a difference to his plants and economy.

The unique selling proposition of the whole scheme is that all operations are to be conducted in the presence of the HDO concerned. They may include schedules and methodology of spraying, sowing, pruning, thinning of fruit or any other action which is beneficial to the farmer.

The department claims to have successfully exhibited to peach farmers the technique and importance of thinning the fruit. The fruit size and quality improved considerably after excess fruit was shed.

It also showed the farmers the benefits of girdling, brushing and dipping of the grape bunches in giberellic acid.

The programme seems to be excellent on paper but much more is required to seek benefits from such actions. Splinter movements like these offer no gain as they fade away with a slightest change in top management.

The department in its cadre has mostly untrained staff who lack modern perspective and new schools of thought. Secondly, the department is devoid of modern tools and implements to run the show. For example, it does not have in its possession spray equipment, pruning scissors, tractors, etc. which are so important for the implementation of this project. The department will be dependent on the farmer to meet its requirements.

Thirdly, it will be very difficult to monitor the undisciplined staff whose commitment level is always doubtful.

Moreover, the department has to find ways and methods which shall highlight its performance in the end and certain yardsticks and comparisons have to be drawn through data to prove and convince the farmer that its package and practices are of benefit to him.

The idea is not to criticise the scheme as on the periphery it seems wonderful but there is many a slip between the cup and the lip. The Horticulture Department breaks the shackles of governmental red tape and hire a non-government organisation (NGO) to evaluate the viability of this programme and how best it can be implemented. By engaging a professional NGO the department shall get a correct picture of its ongoing exercise. It shall wilt away the possibility of a biased report and arrest outdated methods of research and development.

Bihar University to Import 20 Litchi Varieties

RAJENDRA Agriculture University has decided to import 26 varieties of litchi saplings to increase production of the popular seasonal fruit in Bihar.

University sources said that the Central Government had already allowed the import of the saplings from China and Australia. Sources said the juicy fruit could emerge as a major foreign exchange earner in the Country.

Only two varieties of litchi, Shahi and China, are being produced in the state. The yield of the litchi could be increased 20 times by adopting mixed cropping which was also helpful in maintaining the quality of the soil.

Stressing the need for cultivating bee-keeping in litchi orchards, the sources said it could also increase production of honey.

6

Developments in Agricultural Exports

Substantial changes have taken place in the composition and direction of agricultural exports over the past two decades. Although the growth in agricultural trade has continued to lag behind total merchandise trade, there has been a faster growth in processed agricultural products. In addition, while developed countries, because of the much higher effective demand, remain the main import markets for most agricultural products, the rapid rise in income growth in East Asia over the past two decades have seen a rapid rise in demand from that region. This section reviews these trends and their implications for market opportunities for low income agricultural commodity-dependent exporters.

Trade in Processed Agricultural Products Grew Faster than Primary Agricultural Products: Over the last 20 years, the world agricultural trade has witnessed striking structural changes with processed product growing faster than primary products. As a group, exports of processed agricultural products grew at 6 percent annually during 1981-2000 (compared to 3.5 percent for primary products), raising their world market shares from 63 percent in 1981-90 to 68 percent in 1991-2000. Among the processed products growth rates have been exceptionally high (above the average 6 percent) for the processed forms of cereals, fruit, vegetables, pulses, tropical beverages and poultry products. As a result, the share in total world agricultural exports of most of processed items increased significantly.

Primary agricultural exports, on the other hand, grew by 3.5 percent during 1981-2000. Despite this low growth rate for primary products as a whole, there has been considerable growth in some non-traditional primary export items. These include fresh fruit, vegetables, tree nuts, dairy, poultry, spices, with growth rates exceeding 6 percent. The rapid growth of trade in most of these products has been fuelled by changing consumption trends in developed countries. It is generally argued that trade in these products offers opportunities for low income agricultural commodity-dependent exporters that can meet the required standards.

Comparison of country shares in the value of agricultural trade suggests that export shares increased in developed countries in 1991-2000 compared to 1981-1990. This increase is more pronounced in processed products, where the share of developed countries increased from 47 percent in 1981-90 to 52 percent in 1990-2000. The increase was more evident in such products as processed forms of cocoa, coffee, cereals, root crops, sugar and pulses.

While several developing countries diversified away from primary agriculture into processed agricultural products during the last two decades, high dependence on exports of primary agricultural products continued to be a prominent feature in many developing countries. Developing countries' primary commodity exports accounted for 75 percent of total world exports of primary agricultural commodities in 1981-90, and increased to 78 percent in 1991-2000. In contrast, the share of these countries in processed commodities decreased from 53 percent to 48 during the same period. Available statistics for the period 1997-99, show that for 50 developing countries, the top 3 primary agricultural exports account for over 25 percent of total merchandise exports.

For the majority of the agricultural products, the share of developing countries in world agricultural exports declines as the stage of processing increases. This is more evident in the case of coffee and cocoa. In 1981-90, the share of the top 10 cocoa-exporting developing countries in world exports of cocoa beans, cocoa butter, cocoa powder and chocolate were 85 percent, 16 percent, 28 percent and 3 percent respectively. Similarly, while the share in world green coffee exports of the top 10 coffee exporting developing countries remained unchanged at about 65 percent between 1981-90 and 1991-2000, the share of these countries in roasted coffee declined from 7 to 2 percent during the same period.

Agricultural trade among developing countries during 1985-99 grew faster, 12.5 percent per annum, than the agricultural exports to developed countries, which grew at 7.2 percent per annum. Intra-regional agricultural exports as a share of total agricultural trade in Latin America increased from 15.5 percent per annum in 1991-94 to 18 percent in 1995-99. It grew at an average of 19 percent per annum during the 1991-94, but growth slowed down to 1.6 percent in the period 1995-1999. The slow growth in intra-regional trade during the latter period can partly be explained by the economic crisis in the region. In South/East Asia, intra-regional agricultural exports grew by 16 percent per annum during 1991-1994, but showed a negative growth rate during 1995-1999. Nonetheless, their share in total trade increased from 35 percent in 1991-94 to 41 percent in 1995-99. The Asian economic crisis has been the main factor behind the deceleration during 1995-99. Trade flows among the countries of West Asia have shown equally impressive growth rates, raising the share of intra-regional trade from 36 to 42 percent during the periods 1991-94 and 1995-99 respectively.

Africa's intra-regional agricultural exports grew on average of 1.4 percent per annum during 1991-1994 and -1.2 during 1995-99. African agricultural exports to other developing regions increased faster than intra-African exports in 1990-1994 (13 percent to Latin America, 10 percent to West Asia, and 11 percent to South/East Asia). The largest markets for African agricultural exports among developing countries are in South/East Asia with an average annual value of US$ 1.3 billion in 1991-1994, and increase to US$ 1.6 billion in 1995-1999.

This growth in trade among developing countries could in part be related to the proliferation of regional trade agreements (RTAs) among them. In many developing regions, RTAs are seen as a vehicle for promoting and diversifying trade. This is particularly true of those agreements such as the Southern

Common Market (MERCOSUR), the Central American Common Market (CACM), the Caribbean Community (CARICOM), the Association of South East Asian Nations (ASEAN), the Common Market of Eastern and Southern Africa (COMESA), the Southern African Development Community (SADC) and the West African Economic and Monetary Union (UEMOA), which have largely liberalized agricultural trade within their regions. A recent FAO paper assessing the impact of RTAs in Latin America concluded that the implementation of regional trade agreements has been the main factor behind the expansion of agricultural trade in that region.

However, most of the RTAs among poorer developing countries, particularly in Africa, have not been associated with tangible improvement in intra-regional agricultural trade. All such trading efforts have come up against structural and policy obstacles. Difficulties include inadequate international transport and communication facilities and poor information about markets and investment opportunities. Moreover, the absence or inadequacy of a system for standardized packing, grading and quality control systems at regional levels continues to frustrate efforts to expand trade and establish transparent information systems.

Diversification and Adding Value

Recognising the severe difficulties faced by low income agricultural commodity-dependent exporters in the world market, diversification and adding value to agricultural exports has been viewed as the best way to achieve export growth and stability in these countries. There are both horizontal and vertical dimensions to export diversification. Horizontal diversification involves expansion in the export mix into products with higher growth potential, while vertical diversification involves creating additional uses for existing and new products through value-added activities such as processing and marketing.

Although trade in the processed products and high value crops is expanding rapidly, there are a number of obstacles that hinder poorer countries from diversifying into these products and take advantage of the new export opportunities. This has been the result of a combination of demand factors (stagnating world demand, declining prices), unfavourable market access or market entry conditions, competition from subsidized production and supply-side bottlenecks. This section discusses those factors relating to the international agricultural trading system, namely: excessive support and protection to agriculture; preferential trade arrangements; and the changing nature of market structure.

Agricultural Protection And Support

A major impediment to growth in agricultural trade remains to be the high level of protection and support to agriculture in the developed countries. The Uruguay Round initiated the process of opening up new opportunities for export diversification in agriculture, through, *inter alia*; tariffication and across-the-board reductions in most-favoured nation (MFN) tariffs on agricultural products; reduction of trade-distorting support; and the strengthening of trade rules, particularly those on sanitary and phytosanitary measures and technical barriers to trade. Despite these changes, little improvement has been achieved and market access problems persist particularly for processed agricultural products.

Agricultural Tariffs: Agricultural tariffs remain high and complex. Tariffs in the developed countries are generally lower on average, but there are many tariff peaks, particularly on processed products and temperate-zone products such as horticulture, sugar, cereals, dairy products and meat. These tariffs are often complex with several different *ad valorem*, specific and seasonal rates applying to the same products. Bound tariffs are also high in many developing countries, although their applied rates are generally lower.

Recent estimates show that nominal protection rates in the OECD countries have declined somewhat but remain high on the whole. While tariff rate quotas (TRQs) have created some new trading opportunities, consistently low and declining fill rates indicate that there are difficulties in their implementation.

Tariff Escalation: The existence of tariff escalation (tariffs which escalated with the degree of processing) in agricultural markets is regarded as one of the major factors that hinder export growth and diversification in the exporting countries.

Tariff escalation prevails in a large number of agricultural commodity chains. Of 17 commodity chains examined by the Secretariat, 12 suffer from tariff escalation, mostly at the first stage of processing. In developed countries, escalation is particularly pronounced in coffee, cocoa, oilseeds, fruit and hides and skins. In developing countries, however, applied tariffs are often much lower than bound rates and in almost all cases they also show relatively high escalation.

Tariff escalation has been raised as one of the important market access issues in the current WTO negotiations on agriculture. Country proposals addressing tariff escalation are suggesting the adoption of a harmonising reduction formula that reduces higher tariffs by greater amounts, including tariff peaks, and eliminates tariff escalation. A recent FAO study shows that while the adoption of a harmonising formula for reducing tariffs such as the Swiss formula appears to be a better option for reducing tariff escalation than linear methods, it is unlikely that this formula will achieve substantial reduction in tariff escalation in all commodities of export interest to low income agricultural commodity-dependent exporters, particularly tropical products, unless it involves some additional criteria for capping tariff wedges along the processing chain.

Support to Production and Exports: Subsidized production and exports affect both domestic and international markets and exert a negative influence on the diversification of production and exports from commodity exporting countries. Loss of competitiveness caused by subsidized agriculture discourages investments in agriculture and local processing in non-subsidizing countries.

Under the WTO Agreement on Agriculture (AoA), WTO members undertook to reduce both export subsidies and the trade-distorting domestic support. Export subsidies have been reduced somewhat on several products since 1995, but they remain high, particularly for meat and dairy products. In domestic support, some WTO members have shifted to less distorting support measures, but overall support to agriculture remains high. Estimates show that total transfers to agriculture in the OECD countries amounted to US$311 billion in 2001, compared with US$298 billion in 1986-88. The support to producers in these countries as a share of the value of farm output averaged 31 percent for agriculture as a whole and exceeded 40 percent in such products as rice, wheat, sugar, beef and sheep meat. Such support is believed to have stimulated overproduction creating a downward pressure on world prices, thus limiting trading opportunities for non-subsidizing exporters.

International Technical Standards. A major challenge faced by many exporters of agricultural products, particularly in low income countries, is to meet the sanitary, phytosanitary and technical requirements of importing countries. An additional challenge is faced by these countries when new standards are introduced on risk assessment grounds that are stricter than those currently recognized by international standard-setting bodies. Rising consumer concerns in the affluent countries over food quality compound the difficulty of these countries in meeting ever higher standards. The transition to risk-based standard setting has required major changes in legislative, regulatory and administrative practices in most countries all of which have implied significant costs.

The concerns of low income countries in this regard are three-fold: first that they usually lack the technical and financial capacity to fully participate in activities of international standard-setting bodies such as the Codex Alimentarius Commission (CAC) and the International Plant Protection Convention (IPPC). Second, they lack the supply-side capacity to meet the increasingly strict standards being adopted by the developed countries. Third, there are concerns about the scope for the discriminatory use of these measures. A recent FAO study provided a broad indication of this problem based on data on rejections following border inspection in export markets. The import detention list for the United States for the period 1996-97 shows that the majority of detentions and rejections of products coming from developing countries, particularly those of Sub-Saharan Africa, are related to microbiological contaminants and filth rather than to strictly technical considerations. This illustrates the considerable problems that low income countries have in meeting basic food hygiene requirements let alone requirements for which more sophisticated monitoring and testing, and therefore more costly procedures are required, such as limits on pesticide residues and heavy metals.

Preferential Market Access

Despite the current constraints on agricultural trade emanating from protection and support in the developed country markets, preferential trade arrangements involving developing countries have provided some opportunities for these countries to expand and diversify their agricultural exports. Three major forms of trade preferences can be distinguished: the Generalized System of Preferences (GSP), special preferential regimes for groups of developing countries (such as Lome/Cotonou, AGOA, or the Caribbean Basin Initiative), and regional free-trade areas between developed and developing countries. However, the last form differs from the others in that it involves reciprocal rather than unilateral trade preferences. MFN duties are still high for many agricultural products and preference schemes are offering substantial price advantages to some exporters for numerous eligible products. This is specially the case, on a wide scale, for least developed countries, as well as under special preferential arrangements of the EU and the United States for most African, Caribbean, Central American, Andean and Pacific countries.

While many developing countries have been accorded preferences for a large number of products, exceptions to these preferences often relate to agricultural products. For example, the European Union's initiative on "Everything but Arms" (EBA) offers free market access to LDC products, with less than 5 percent of pre-EBA exports left facing a tariff barrier. A recent study by UNCTAD, however, shows that the impact of this initiative will be a relatively small increase, in the medium term, in exports from the LDCs, as 70 percent of the potential positive trade effects would have come from free access for sugar, rice and beef, which has been deferred until 2006. The African Growth and Opportunity Act has also started to open up markets in the United States for African countries.

Although trade preferences are of significant benefits to a number of countries, in general they have been, to a large extent, underutilised. Part of the reason for this underutilisation is due to, in many cases, lack of security of access, insufficient product coverage, excessively stringent rules of origin and often a lack of awareness of the preferences available. These factors are in addition to supply-side constraints faced by some countries, in particular the least developed countries

Market Structure and rhe Distribution Of Gains

Government barriers to trade are important, but as they are reduced other factors impeding market entry, particularly those that spring from restrictive business practices, come to the forefront.

The agricultural commodity chains, particularly those of high value crops and processed products, are increasingly dominated by a few multinational enterprises (MNEs) and distribution companies. This dominance of large companies in the world agricultural commodity markets is not only through their actions on international markets, but also through their direct and increasing influence on what is produced, and how. While these changes are opening up unprecedented opportunities for some producers and exporters, benefiting from these developments and avoiding their negative impacts requires much greater business skills than before on the part of Governments and small-scale entrepreneurs, particularly in low-income countries.

Changes in agricultural commodity chains, as a result of the increasing dominance of large companies, can be reviewed at three levels, namely, the international markets, exporting developing countries, and the retail markets of importing countries. At the international level, agricultural trade continues to be concentrated in a few vertically integrated MNEs. In 1996, for example, five MNEs accounted for about 50 percent of world trade in green coffee, up from about 37 percent in 1980; and four coffee roasters accounted for 50 percent of the roasted coffee market. In cocoa, the number of trading houses in London has been reduced from 30 in 1980 to around 10 in 1999. Similarly, the six largest chocolate manufacturers account for 50 percent of world chocolate sales. For vegetable oils, following a series of mergers and acquisitions in the 1990s, a small number of vertically integrated MNEs now dominate the production, distribution and international trade of both oilseeds and oils. In grains, a few big companies are transforming themselves, through mergers and consolidations, into vertically integrated businesses in trading, storage, processing and milling.

In exporting developing countries, following the deregulation and the disappearance of many marketing boards, large companies with warehousing and shipping facilities in the producing countries are able to exploit their financial and logistical advantages, even buying the produce directly from the farmer. Intensified competition favours those with access to cheaper finance and good logistics. Being big provides advantages on both accounts.

At the retail level, supermarkets have also being growing rapidly in both developed and developing countries. For example, there has been rapid consolidation and multinationalisation of the supermarket sector in Latin America, where the share of supermarkets in food retailing went from 15-20 percent in 1990 to 60 percent in 2000. Given their dominance in the market, supermarkets have a significant impact on production, distribution and trade. To simplify operations, most of these supermarkets are increasingly cutting the number of their suppliers, and as a result those left out will have to sell in a shrinking and low-priced market or find a niche market. In addition, supermarkets have significant impact on production requirements in terms of quality and food safety, product variety and delivery. The Food Business Forum (CIES), through its association of the world's 200 largest supermarkets and their suppliers (with a combined sales of US$2 800 billion), decided earlier this year on new mandatory standards for their suppliers. While this move may help in reducing the number of different standards, it may overshadow whatever Governments decide in the context of the WTO agreements.

The stringent private standards set by supermarkets are very often a reflection of consumers' concerns. Consumers of agricultural commodities are increasingly choosy with respect to the quality of the commodities they buy, the timeliness of delivery, origin and traceability, and the environmental and social conditions under which they have been produced. The use of specialised channels such as fair-trade labelling and other socially or environmentally responsible market practices, and taking advantage of niche market such as that for organic food, may help in facilitating market entry and can improve earnings of producers. In many countries, consumers are willing to pay a price premium (of more than 20 percent) for organic foods, particularly fruit and vegetables.

There is evidence that the increasing market concentration along agricultural commodity chains has been associated with a declining share of producing countries in the revenues generated from the sale of these products in global markets. Recent studies show a widening gap between world commodity prices and consumer prices in industrialized countries. Growers' price represents very low shares of the final price of the product, ranging from 4-8 percent for raw cotton and tobacco to 11– 24 percent for jute and coffee.

Similar patterns were also observed in fresh fruit and vegetables. The International Coffee Organisation (ICO) has reported that in the early 1990s export earnings by coffee producing countries were about US$10-12 billion and the value of retail sales of coffee, largely in the developed countries, was about US$30 billion. However, in coffee year 2000/01, producing countries received only US$5.8 billion of the value of retail sales of more than US$60 billion.

Studies on commodity chains involving exports of fresh vegetables and fruit from Africa to the UK have shown that returns to the participant in the market are highly concentrated at the end of the chain in the importing countries. As shown in the figures below, returns to producer, exporter and packaging of fresh vegetables exported from Kenya and mangetout from Zimbabwe amounted to only 27 and 23 percent of the final price of these products respectively.

Conclusions: While recent market trends and policy developments are opening up unprecedented opportunities for some producers and exporters, they also pose formidable challenges for low income agricultural commodity-dependent exporters. Benefiting from such changes and avoiding their negative impacts require considerable enhancement of productive capacities, the upgrading of food quality and safety assurance systems, and much greater business skills than before on the part of entrepreneurs. While these supply-side measures are of fundamental importance for taking advantage of the new trading opportunities, two major problems relating to the international trading system require much more attention by the international community: the continuation of trade distorting agricultural support policies, and the rising market entry obstacles as a result of increasing market concentration along the agricultural commodity chains.

If all countries are to benefit from global trade, distortions to global agricultural markets due to government interventions must be reduced. Significant improvements in market access will require, apart from a sharp and meaningful reduction of subsidies, much deeper cuts in tariffs and tariff escalation. In so doing, there would be a need to take into account transitional problems likely to be faced by those countries, particularly low-income countries, whose trade is dependent on preferential access, and a possible temporary rise in food prices. There also appears to be a need to deal with issues relating to market structure and competition in the market chain of various agricultural commodities, which determine the distribution of gains for different participants in particular commodity sectors. Much more research is needed on the changing nature of market structure and supply chains and their implications for agricultural development and food security, particularly for low income agricultural commodity-dependent exporters.

The Committee may wish to consider the issues raised in this document and to provide guidance to the Secretariat on further research and analysis.

Food Policy and Nutrition Division

Roots and tubers belong to the class of foods that basically provide energy in the human diet in the form of carbohydrates.

The terms refer to any growing plant that stores edible material in subterranean root, corm or tuber. The development of root crops in the tropics was accelerated by the introduction of gari-processing technology into West Africa and by the promotion of cassava as a famine reserve by several colonial governments, such as the Dutch in Java and the British in West Africa and India. By 1880 the tapioca trade was well established in Malaya and by the turn of the twentieth century the production and trade of cassava products, especially starch, had been established by the Dutch in Java and by the French in Madagascar.

A further reason for the spread was the fact that during tribal warfare and invasions, the invader could not destroy or remove the food reserve, which could be kept conveniently under the ground, thus giving added food security to the population.

Historically, very little attention has been paid to root crops by policy-makers and researchers as most of their efforts have been concentrated on cash crops or the more familiar grains. Root crops were regarded as food mainly for the poor, and have played a very minor role in international trade. This misconception has lingered for so long because of the lack of appreciation of the number of people who depend on these root crops, and the number of lives that have been saved during famine or disasters by root crops.

It was cassava that saved the Rwanda-Burundi kingdoms in 1943 when potato blight destroyed all their production, and cassava also fed the Biafrans during the Biafran war in Nigeria in 1966-69.

As far back as 1844 Rev. John Graham has this to say about the potato:

"Oh! There's not in the wide world a race that can beat us,
From Canada's cold hills to sultry Japan,
While we fatten and feast on the smiling potatoes
Of Erin's green valleys so friendly to man."

There is an old saying of the Palananans of Micronesia where taro is the basis of the staple food that:

"the taro swamp is the mother of life" (Kahn, 1985).

The fact that these root crops are mainly starchy has led to the disparagement of their protein content, which is low compared to cereals. However, considering the quantities of root crops consumed a day, their protein contribution is often significant. In addition, root crops contain an appreciable amount of vitamins and minerals and may have a competitive production advantage in terms of energy yield per hectare over cereals produced in ecologically difficult conditions.

During the years 1980-87, the average rate of growth in food production (2.6 percent) in many countries of developing market economies, particularly Africa, has been either falling behind or barely keeping pace with the average annual rate of growth in population (about 3 percent), owing to land shortages and lack of foreign exchange to purchase agricultural inputs such as fertilizers, insecticides and m achinery. Droughts, floods and other natural and induced calamities have contributed substantially to reduced food supplies. These fooddeficit countries are now making substantial efforts to improve this situation, but their efforts are directed mainly toward improving production of staple cereals and increasing the importation of cereals by cash purchases or as food aid, thereby widening further the gap between local food production and food requirements.

In their present state under subsistence farming, the yield of many root crops is very low, but their genetic potential for producing increased yields is high and has not yet been fully exploited. In addition some root crops are highly adaptable, producing reasonable yields from marginal lands with highly erratic rainfalls. Crops such as cassava can serve as a valuable asset for household food security for subsistence populations in times of drought and under other unfavourable ecological conditions.

The purpose of this book is to review the value of roots and tubers in the human diet and to assess their contribution to the nutritional welfare and food security of people living in less developed countries. It is hoped that it will also help propagate knowledge about these crops and stimulate research for their genetic improvement with a view to increasing their production and utilization.

In its Food and Nutrition Papers and Series, the Food and Agriculture Organization of the United Nations (FAO) has already published books on five important food sources, namely Rice and rice diets, Maize and maize diets, Milk and milk products in human nutrition, Wheat in human nutrition and Legumes in human nutrition. The present study has been conceived on a similar pattern. It will summarize current knowledge about production, consumption, nutritive value, processing and cooking of roots and tubers and their contribution to hum an diets. The study includes all important root and tuber crops: cassava, yam, sweet potato, potato and aroids as well as two other starchy staples, banana and plantain. Although much of the research and development work on potato is done in temperate zones, the potato is included in the book because of its great potential for expansion into the tropics. Plantain and banana are also important starchy staples in many tropical countries.

In 1988 FAO published a study entitled Root and tuber crops, plantains and bananas in developing countries: challenges and opportunities, which provides a comprehensive review of the global production and consumption of these crops. Additional FAO studies in this field focus on the utilization and processing of roots, tubers, plantains and bananas. The present study provides a more detailed analysis of the role of these crops in human nutrition and is an essential supplement to the information provided in the earlier publications.

Origins and Distribution

The suggested origins of root and tuber crops are illustrated. These crops were dispersed by the Portuguese during their voyages for slaves, by both the Portuguese and Spanish in their missionary journeys, and by Arab traders. The genus Dioscorea (a variety of yam) has a wider diversity of origin with different species adapted to different ecosystems. D. trifida is indigenous to tropical America; D. rotundata, D. cayenensis, D. bulbifera and D. dumetorum are native to West Africa; D. alata, D. esculenta and D. opposita are indigenous to South Asia. D. opposita and D. japonica have their centre of origin in China.

Yams are the only root crops in which the Asian and African species developed independently of each other. Exchange of species was due to the influence of Portuguese explorers. They learned of the value of D. alata from the Indian and Malayan seafarers who used it on their ships on long voyages because it stored well and had antiscorbutic properties. The Portuguese soon adopted it and introduced it into Elmina and Sao Tome in West Africa. Subsequently, through the Atlantic slave trade, the Portuguese carried the African species D. rotundata and D. cayenensis and the Asian species D. alata to the Caribbean where they became important staple foods (Coursey, 1976). According to Coursey (1967), D. alata seems to have arisen from the wild relatives, D. hamiltoni and D. persimilis in the north and central pans of the southeast Asian peninsula, probably Burma or Assam. So also D. esculenta while D. hispida, D. pentaphylla and D. bulbifera origina ted from an IndoMalayan centre. D. rotundata is of African origin, where it is known as "water yam", indicating that it was brought across the water or sea. D. rotundata is the most important African yam, especially in the forest zone, and is probably a hybrid of the other African yam, D. cayenensis, which is a savannah species. In West Africa it is grown in the roots and tubers belt, which extends 15°N and 15°S of the equator (Coursey, 1976; Okigbo, 1978; Nweke, 1981).

Little is known about the origin of new world yams. They were of secondary importance in the pre-Colombian era. D. trifida, an Amerindian domesticate, appears to have originated on the borders of Brazil and Guyana, followed by a dispersion through the Caribbean (Ayensu and Coursey, 1972). Yams were taken to the Americas through precolonial Portuguese and Spanish expansion that began around 500 years ago. Historical records of D. alata in West Africa and of African yams in the Americas date back to the sixteenth century (Coursey, 1967).

Sweet potato, which originated in the Yucatan peninsula in Latin America, seems to be the most widely dispersed root crop. It is adaptable and can grow under many different ecological conditions. It has a shorter growth period than most other root crops (three to five months) and shows no marked seasonality: under suitable climatic conditions it can be grown all the year round. Adverse weather conditions rarely cause a complete crop loss. Hence sweet potatoes are planted as an "insurance crop", combined in mixed cropping with grains like rice in Southeast Asia, and with other root crops like cocoyam and yam in Oceania. It is a popular plant in the Philippines and in Japan because of its prostrate habit, which makes it resistant to damage by high winds such as hurricanes and typhoons (Wilson, 1977). Sweet potatoes have been cultivated since about 3 000 B.C. and were an important food for the Mayans in Central America and the Peruvians in the Andes. From ethno-historical records of Colombia, the reports of Spanish explorers and missionaries in Mexico and Peru, and of the Portuguese in Brazil, it is clear that sweet potato was common throughout the American tropics before 1492. The plant was further dispersed by Iberian and Portuguese explorers to the Pacific area in the sixteenth century. The Portuguese explorers subsequently transferred West Indian clones, grown in the western Mediterranean area, to Africa, India and the East Indies. Spanish traders also took sweet potatoes from Mexico to Manila. Later the sweet potato reached New Guinea and the eastern Pacific Islands and extended into China and Japan. It is now grown extensively in a wide range of environments between latitude 40°N and 40°S and from sea level up to an altitude of 2 300 metres.

The distribution of potato is also extensive. It originated in the high Andes of South America where it was adapted to the cold climate and short days prevailing in those latitudes. Wild cultivars are still found on elevated regions extending from the southwestern part of the United States of America to the southern part of South America, and more especially at high altitudes in Bolivia and Peru and in the coastal regions and nearby islands of southern Chile (Simmonds, 1976). When the original potato was first introduced to Europe it remained a botanical curiosity for more than a century and it did not flourish until a variety adapted to the longer day was evolved.

Spanish sailors introduced the potato to Spain as early as 1573. It was probably introduced into England by English seamen from captured Spanish ships around 1590. From Spain, potatoes spread throughout continental Europe; from England, they were dispersed throughout Great Britain to parts of northern

Europe. By 1600, potatoes were sent from Spain to Italy and from there to Germany and in the same year they reached France.

Potatoes reached most other parts of the world through European colonial activities. North America received potatoes from England in 1621, British missionaries took potatoes to Asia in the seventeenth century and Belgium missionaries carried them to the Congo in the nineteenth century. The potato was brought to India in the sixteenth century by Portuguese traders and within about 200 years it had spread all over India. It was taken to Bhutan, Nepal and Sikkin from India. In Africa, introduction of potatoes followed colonization. Possibly, its antiscorbutic properties persuaded seafarers to stock it in their ship's store and to encourage people to grow it wherever they visited.

Like sweet potato, potato also exhibits a growing period of about four months, shorter than many other root crops. The indigenous South American cultivars will develop tubers at longer day lengths than other root crops and numerous cultivars will even tolerate the extreme day length of 24 hours of the polar summer (Kay, 1973). So the spread was very easy.

A typical example of a root crop that can tolerate drought and poor husbandry is cassava. Cassava originated in tropical America but the precise area of its origin is not known. The two probable areas suggested are the Mexican and Central American area or northern South America. It was first introduced into the Congo basin as early as 1558 by the Portuguese. It then spread rapidly through Angola, Zaire, Congo and Gabon and later to West Africa. There was a separate introduction to the east coast of Africa and to Madagascar in the eighteenth century by Portuguese and Arab traders, after which it rapidly became a dietary staple throughout many lowland tropical areas (Jones, 1959). The cultivation of cassava in Africa increased during the nineteenth and twentieth centuries as a result of encouragement by administrative authorities, who recognized its value as a famine relief crop. According to Kahn (1985), after the First World War, the farmers of RuandaUrundi, now the independent nations of Rwanda and Burundi, at first refused to take advice from the Belgians to plant cassava, because they had enough potatoes. But in 1924 the Belgians issued a strict order to grow cassava and recruited 60 000 porters to carry 5 000 tonnes of cassava stakes around the region for planting, so the farmers finally accepted.

Cassava was taken to India by the Portuguese in the seventeenth century. In about 1850, it was transported directly from Brazil to Java, Singapore and Malaya. Cassava was introduced to the South Pacific territories during the first half of the nineteenth century by missionaries and travellers but its importance varies from island to island. At present, cassava is grown throughout tropical and subtropical areas approximating 30°N and 30°S of the equator and up to an altitude of 1 500 metres.

The spread of root crops was facilitated by their ability to thrive under varied tropical conditions. Their level of water tolerance varies considerably, ranging from the waterlogged conditions required for taro to the drought tolerance and minimal water supplies needed for cassava once it has been established (Wilson, 1977). It was the requirement of flooded conditions for taro, Colocasia esculenta, that convinced anthropologists that these yams were the first irrigated crops, and that the ancient "rice" terraces of Asia were originally constructed for them (Plucknett et al., 1970). Tannia (Xanthosoma sagittifolium) on the other hand cannot tolerate being waterlogged (Onwueme, 1978).

Xanthosoma, or new cocoyam, had its origin in South America and the Caribbean. The Spanish and Portuguese introduced it to Europe and were also responsible for spreading it to Asia. It moved from the Caribbean in the late nineteenth century, first to Sierra Leone and then to Ghana. In West Africa, Xanthosoma is more important than Colocasia, being popular for its corm, cormels, leaves and young stems. Although Xanthosoma is relatively new to the Pacific region, it has spread rapidly and widely, becoming quite an important crop in many of the islands. It is also widely cultivated in Puerto Rico, the Dominican Republic and Cuba and is important along the coastal mountains of South America, in the Amazon basin and in Central America.

Colocasia originated in India and Southeast Asia. About 2000 years ago it spread to Egypt and thence to Europe (Plucknett et al., 1970). Subsequently it was taken from Spain to tropical America and then to West Africa. It was used in feeding slaves and was transferred to the West Indies with the slave trade (Coursey, 1968). In order to distinguish it from the newer species, Xanthosoma, Colocasia was referred to as "old yam" in West Africa whereas Xanthosoma is called "new yam". Colocasia is a staple food in many islands of the South Pacific, such as Tonga and Western Samoa, and in Papua New Guinea. Colocasia and Xanthosoma will tolerate shade conditions and so they are often planted under permanent plantations like banana, coconut, citrus, oil palm and especially cocoa. Therefore they are sometimes collectively referred to as cocoyams.

The banana is believed to have originated in Southeast Asia, having been cultivated in South India around 500 BC. From here it was distributed to Malaya through Madagascar and linen moved eastward across the Pacific to Japan and Samoa in the mid-Pacific at about AD 1000. It was probably introduced to East Africa around AD 500 and had become well established in West Africa by AD 1400. It finally arrived in the Caribbean and Latin America soon after AD 1500 (Simmonds, 1962; 1966; 1976). By the end of the eleventh century, banana had spread widely throughout the tropics. In South America it was found as far south as Bolivia and was cultivated in most of Brazil. In Africa banana growing extended from the Sahara to Tanzania in the east and from Cote d'Ivoire through the Congo to Zaire in the west and central areas.

New Frontiers for Tropical Root Crops

In this chapter agro-industrial possibilities for the expanded utilization of root crops are reviewed. The success of efforts made to increase the production of tropical root crops and to promote their use as food will depend on market demand. Farmers will not be encouraged to produce a marketable excess if this leads to glut, spoilage, and low prices. Policy makers should not only promote policies to increase consumption of root crops as human foods and as animal feeds, but should also support research that will extend the utilization of these root crops. Efforts should be made to promote new technologies, appropriate for use by the rural population, to produce a variety of processed foods from root crops. This strategy will generate employment and improve incomes in rural areas. If demand is stimulated farmers will be encouraged to produce more root crops which can be converted to animal feed and industrial uses. Demand can be stimulated by development in three main areas:

- commercial dehydration of the root crops to produce flakes and flour that can be made into other food products;
- use of root crops as sources of industrial raw materials; and
- use of root crops as animal feed.

Commercial Dehydration of Root Crops

The peeled root is rinsed to remove excess starch, then cut into slices, blanched, blended to a puree and dried. Peeling can be effected by immersion in 10 percent lye solution or by steaming at high temperatures (150°C) for short periods.

The drier, which may be a heat exchanger or drum drier, can be fired by agricultural wastes such as coconut husks which are abundant and cheap in Southeast Asia. This reduces the moisture content from 70 percent to 12 percent. Improved preservation of root crops would increase their availability and reduce post-harvest wastage. Dried products require less storage space and have a longer shelf-life. They can be quickly reconstituted and prepared for eating, a factor of particular importance to urban consumers. Composite flour incorporating cocoyam has been used in extruded products such as noodles and macaroni. Similar processes could be used in the production of flour products from other root crops.

Processing will greatly increase the utilization of root crops. The flour can be used as a component of multimix baby foods and in composite flour for making bread. Research and development work on composite flour using root crops and other local products has advanced considerably in Colombia. Based on initial research in 1971-72, it was concluded that while rice and maize flours are preferable for use as non-wheat components in composite flours, cassava flour and starch also have good technical possibilities. The pilot work demonstrated that the production of bread from wheat flour diluted up to 30 percent with non-wheat components is possible on a commercial scale. But the large-scale introduction of such flours requires a concerted effort by both the public and private sectors to ensure the wide-spread availability at attractive prices of non-wheat raw materials. Expanded cassava production and lower prices are required if composite flour is to be economically attractive to millers and consumers (Goering, 1979).

Bread baked with composite flour from local resources would reduce the foreign exchange cost of imported wheat. This cost is particularly high in the Philippines, where a processing plant has been set up to convert 5 000 kg of fresh sweet potato into flour every day. The bread from this flour will contain more calories and a higher content of vitamin A and lysine than wheat bread and will conserve foreign exchange. If it can be marketed at a reduced cost it may help to improve the nutritional status of the population. Taylor (1982) estimated that cash benefits to farmers producing raw material for this plant could substantially improve with a guaranteed market. Since the market and the financing of the crop is guaranteed, the promotion of sweet potato as a cash crop will be more easily accepted.

Fresh root crops are rarely exported in appreciable quantities because of their high water content and perishability. Cocoyam is exported in small amounts from Fiji, Western Samoa, Tonga and Cooke Island to the United States of America, New Zealand and Australia for the Polynesian and Melanesian immigrants. Yams are also exported from Latin America and Africa for immigrants in Europe, but quantities are small and prices are high.

Bibliography

Adams H.: *Hot Pepper Improvement*, St. Michael, CARDI, 1997.

Agarwal P. K.: *Improvement of Citrus*, New Delhi, Malhotra Publishing House, 1993.

Alexander M. P. and Ganeshan S.: *Pollen Storage*, New Delhi, Malhotra Publishing House, 1993.

Andrews J.: *The Domesticated Capsicums*, University of Texas Press. 1995.

Andrews L.: *Citrus Production - Orange*, St. Augustine, Trinidad and Tobago, 1990

————: *Production of Minor Crops Sapodilla, Tamarind, Guava and Genip*, Port of Spain, Trinidad and Tobago, 1996.

Arguello D. S.: *Production, Post-harvest Management and Exportation of Tropical Fruits in Costa Rica*, Wageningen, Technical Centre for Agricultural and Rural Cooperation, 1992.

Ashworth S.: *Seed to Seed*, Decorah, Seed Savers Publications, 1991.

Barbeau G.: *Tropical Fruits in Nicaragua*, Managua, Nicaragua Ministerio de Desarrollo Agropecuario, Agraria, 1990.

Batlle I. and Tous J.: *Carob Tree (Ceratonia siliqua L.)*. Rome, International Plant Genetic Resources Institute, 1997.

Berger, P.L.: *Pyramids of Sacrifice: Political Ethics and Social Change*, New York, Basic Books, 1974.

————: *Pyramids of Sacrifice: Political Ethics and Social Change*, New York, Basic Books, 1974.

Brooks, D.: *Water: Local-Level Management*, Ottawa, International Development Research Centre, 2002.

Bunnik J. S. C.: *Fresh Fruits and Vegetable: a Survey on the Netherlands and other Major Markets in the European*

Community, Netherlands, Centre for the Promotion of Imports from Developing Countries, 1990.

Burton W.G.: *The Potato*, Holland, H. Veenman & Zonen N.V., 1966.

Cao Van P.: *An Integrated Approach for the Production and Processing of Minor Fruits in Martinique (FWI)*, Port of Spain, Trinidad and Tobago, 1996.

Cardona M. J. G., Carvajal S. L. B., Salinas D. G. C. and Isaza R. G. B. : *Proceedings of the International Seminar on Plantain Production, Quindio, Colombia,* Quindio, Corporation Colombiana de Investigation Agropecuaria, 1998.

Chadha K. L. and Pareek O. P.: *Advances in Horticulture: Fruit Crops,* New Delhi, Malhotra Publishing House, 1993.

Chambers, R.: *Rural Development: Putting the Last First*, London, Longman, 1983.

Collymore L.: *Fruit Production in Barbados*, Port of Spain, Trinidad and Tobago, 1996.

Coste R.: *Coffee: the Plant and the Product*, London, MacMillan, 1992.

Crucefix D.: *Avocado Variety Selection for Export Development,* Roseau, CARDI, 1996.

Cull Brian & Lindsay Pax: *Fruit Growing in Warm Climates for Commercial Growers & Home Gardeners*, Australia, Reed Books, 1995.

Currah L. and Proctor F. J.: *Onions in Tropical Regions*, Kent, Natural Resources Institute, 1990.

Daniells J.: *Illustrated Guide to the Identification of Banana varieties in the South Pacific*, Canberra, ACIAR, 1995.

Degras L.: *Yam: a Tropical Root Crop*, Wageningen, CTA/ MacMillan, 1993.

Dhatt A. S. and Singh Z.: *Propagation and Rootstocks of Citrus*, New Delhi, Malhotra Publishers, 1993.

Diederichsen A.: *Coriander (Coriandrum sativum L.)*. Rome, International Plant Genetic Resources Institute, 1996.

Doijode S. D.: *Seed Germination in Fruits*, New Delhi, Malhotra Publishers, 1993.

Dudley, E.: *The Critical Villager: Beyond Community Participation*, London, Routledge, 1993.

Featherly H. I.: *Taxonomic Terminology of the Higher Plants*, USA, Iowa State College Press, 1954.

Ferentinos L.: *Proceeding of the Sustainable Taro Culture for the Pacific Conference*, Honolulu, HITAHR, 1993.

Forde S: *Proceedings of CARDI / CTA Workshop on Marketability of Caribbean Minor Fruits*, Port of Spain, Trinidad and Tobago, 1996.

Georges, S.: *The Debt Boomerang: How Third World Debt Harms Us All*, Boulder, Westview Press, 1992.

Glowinski Louis: *The Complete Book of Fruit Growing in Australia*, Australia, Lothian Publishing Company Pty. Ltd., 1991.

Godden G.: *Growing Citrus Trees*, Australia, Lothian Publishing Company Pvt. Ltd., 1988.

Gowen S.: *Banana and Plantains*, London, Chapman Hall, 1996.

Green S.K. and Kim J. S.: *Sources of Resistance to Viruses of Pepper (Capsicum spp.): a Catalogue*, Taipei, AVRDC, 1994.

Gunjate R. T. and Tawde A. B.: *Propagation and Rootstocks of Mango*, New Delhi, Malhotra, 1993.

Hartley W.: *A Checklist of Economic Plants in Australia*, Melbourne, C.S.I.R.O., 1979.

Harwood, R. R.: *Marshalling Technology for Development: Proceedings of a Symposium*, Washington, DC, National Academy Press, 1995.

Herklots G. A. C.: *Vegetables in South East Asia*, London, George Allen & Unwin Ltd., 1972.

Hessayon D. G. Dr.: *The Vegetable Expert*, England, PBI, Publications, 1985.

Huaman Z.: *Descriptors for Sweet Potato*, Rome, International Board for Plant Genetic Resources, 1991.

Hurst Jacqui & Rutherford Lyn.: *A Gourmet's Book of Mushrooms & Truffles*, Sydney, Golden Press Pvt. Ltd., 1991.

Jacquat Christiane: *Plants from the Markets of Thailand*, Bangkok, Duang Kamol, 1990.

Jeffers P.: *Evaluation of Four Onion Varieties in Montserrat*; Plymouth, CARDI, 1992.

Kenridge K. C. and Hardy B.: *Biology and Agronomy of Forage Arachis*, Cali, International Centre for Tropical Agriculture, 1994.

Kroll R.: *Cut Flowers*, Wageningen, CTA, 1995.

Kunelius T.: *Annual Ryegrasses in Atlantic Canada*, Ottawa, Agriculture Canada, 1991.

Lofgren, H., Richards, A.: *Food Security, Poverty, and Economic Policy in the Middle East and North Africa*, Washington, DC, International Food Policy Research Institute, 2003.

Mabberley D. J.: *The Plant-Book : a Portable Dictionary of the Vascular Plants*, Cambridge, Cambridge University Press, 1997.

Madulid Domingo A.: *A Pictorial Cyclopedia of Philippine Ornamental Plants*, Philippines, Makati Metro Manila, 1995.

Malins A.: *Postharvest Handling of Pineapple and Mango*, Port of Spain, Trinidad and Tobago, 1992.

Mannetje, L. T. & Jones, R. M.: *Plant Resources of South-East Asia,* Wageningen, Pudoc Scientific Publishers, 1992.

Matsuoka H.: *Cultivation of Panicum Genetic Resources for Evaluation of Characteristics*, Tokyo, JICA, 1997.

Mc Donald F.: *Plant Tissue Culture Manual for Yam, Cassava, Sweet Potato, Dasheen (taro) and Tannia (cocoyam)*, Roseau, Dominica and Cave Hill, 1993.

Mellor, J. W.: *The New Economics of Growth*, Ithaca, Cornell University Press, 1976.

Miller William: *Dictionary of English Plant Names*, London, John Murray, 1884.

Mitra S.: *Postharvest Physiology and Storage of Tropical and Subtropical Fruits*, Oxon, CABI, 1997.

Morris M. L. : *Maize Seed Industries in Developing Countries*, Colorado, Lynne Rienner, 1998.

Morton Julia F.: *Fruits of Warm Climates*, Miami, Julia F. Morton Publisher, 1987.

Murali T. P. and Duncan E. J.: *In Vitro Propagation of Banana through Aseptic Manipulation of Male Inflorescences*, Port of Spain, NIHERST, 1990.

Nabhan G. P.: *Wild Phaseolus Ecogeography in the Sierra Madre Occidental, Mexico: Areographic Techniques for Targeting and Conserving Species Diversity*, Rome, International Board for Plant Genetic Resources, 1990.

Nijdam J. & De Jong A.: *Elsevier's Dictionary of Horticulture in Nine Languages*, Elsevier Scientific Publishing Co. Amsterdam; New York.

Oka H. I.: *Origin of Cultivated Rice*, Elsevier, Japan Scientific Societies Press, 1988.

Oldham P.: *Cost of Production of Major Tree Crops in Dominica*, Roseau, Ministry of Agriculture, 1991.

Ou Yang Jue Ya: *Comparative Study of Mandarin and Cantonese*, Beijing, China Social Sciences Publishing House, 1993.

Percival John: *The Wheat Plant - A Monograph*, London, Duckworth & Co., 1921.

Pilgrim R.: *Post Harvest Handling of Minor Exotics*, St. George's, Grenada, 1996.

Ragone D.: *Breadfruit: Artocarpus Altilis (Parkinson) Fosberg*, Rome, International Plant Genetic Resources Institute, 1997.

Rehm Sigmund: *Multilingual Dictionary of Agronomic Plants*, Boston, Kluwer Academic Publishers, 1994.

Roy S. K.: *Research on Multipurpose Tree Species in Asia: In Vitro Clonal Propagation of Artocarpus Heterophyllus*, Bangkok, Winrock International Institute for Agricultural Development, 1991.

Singh H. P. and Chadha K. L.: *Genetic Resources of Citrus*, New Delhi, Malhotra Publishing House, 1993.

Sinha G. C., Reddy Y. T. N. and Singh G.: *Propagation and Rootstocks in Guava*, New Delhi, Malhotra Publishers, 1993.

Sperling L. and Berkowitz P. *Partners in selection: bean breeders and women bean experts in Rwanda*. Washington, USA: CGIAR, 1994.

Sreekumar V., Indrasenan G and Mammen G.: *Studies of the Quantitative and Qualitative Attributes of Ginger Cultivars*, Calicut, Kasaragod, 1990.

Stover R. H. and Simmonds N. W.: *Bananas*, United Kingdom: Longman Scientific and Technical, 1991.

Thavarasook C.: *Proceedings of the Mungbean Meeting 90, Chiang Mai, Thailand,* Bangkok, Tropical Agricultural Research Centre, 1991.

Thomas E.: *Fruit Production in St. Kitts and Nevis*, Port of Spain, IICA, 1996.

Vargas E. M., Macaya G., Baudoin J. P. and Rocha O. J.: *Variation in the Content of Phaseolin in Wild Populations of Lima Beans (Phaseolus lunatus L.) in the Central Valley of Costa Rica*. Plant Genetic Resources Newsletter, 2000.

Webb M.: *Weed Problems and their Control in Rice, Papaya, Citrus and Sugar cane: a Report to the Belize Plant Protection Service*, United Kingdom, Natural Resources Institute, 1993.

Whealy K.: *The Garden Seed Inventory*, Decorah, Seed Saver Publications, 1988.

Whitwell A.: *Dominica Orchard Crop Management and Research Project: Pest and Disease Management,* United Kingdom, Natural Resources Institute, 1991.

Woolfe Jennifer A.: *The Potato in the Human Diet*, Cambridge, Cambridge University Press, 1989.

Yoshida T.: *Cultivation of Citrus Genetic Resources for Evaluation of Characteristics*, Tokyo, JICA, 1996.

Index

A

B

C

D

E

F

G

H

I

R

S

T

U

V

W

❑❑❑